AF598370
WITHDRAWN

Methods in Molecular Biology

Series Editor
John M. Walker
School of Life and Medical Sciences
University of Hertfordshire
Hatfield, Hertfordshire, AL10 9AB, UK

For further volumes:
http://www.springer.com/series/7651

Imaging Mass Spectrometry

Methods and Protocols

Edited by

Laura M. Cole

The Center for Mass Spectrometry Imaging, Biomolecular Science Research Center, Sheffield Hallam University, Sheffield, South Yorkshire, UK

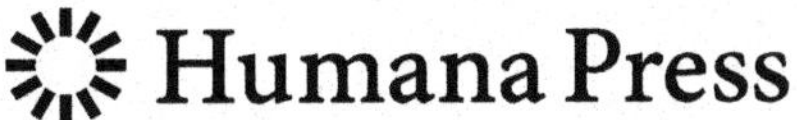

Editor
Laura M. Cole
The Center for Mass Spectrometry Imaging
Biomolecular Science Research Center
Sheffield Hallam University
Sheffield, South Yorkshire, UK

ISSN 1064-3745 ISSN 1940-6029 (electronic)
Methods in Molecular Biology
ISBN 978-1-4939-7050-6 ISBN 978-1-4939-7051-3 (eBook)
DOI 10.1007/978-1-4939-7051-3

Library of Congress Control Number: 2017939579

Printed on acid-free paper

This Humana Press imprint is published by Springer Nature
The registered company is Springer Science+Business Media LLC
The registered company address is: 233 Spring Street, New York, NY 10013, U.S.A.

Preface

The concept of mass spectrometry imaging (MSI) encompasses a range of powerful analytical techniques, all with the capability to explore the molecular signatures and chemical biology of numerous biological samples. MSI not only provides a molecular snapshot of the tissue sample being analyzed but also allows preservation of the spatial distribution of a target species of interest.

From lipids and metabolites to proteomic macromolecules, MSI can track molecular profiles which can be subsequently aligned to correlate specific anatomical features of the tissue under study. Moreover, it enables the mapping of ion profiles, which could lead to the prediction of functional partners that link biochemical pathways and whole disease networks.

There has been a meteoric advancement over the last 20 years in MSI technology and methodologies, and subsequently this unique process has been adopted as a worldwide complementary research tool featuring in numerous research projects. This protocol series aims to address the various techniques, novel applications of MSI, and its role as a discovery tool in the field of proteomics, lipidomics, and metabolomics. The authors demonstrate through their areas of expertise how MSI can be applied to many areas of research including clinical pathology, translational medicine, toxicology, biomarker and response studies along with the potential integration of MSI into forensic workflows.

In addition to the traditional protocol format of this book, other authors discuss novel image processing, future models of disease, and ethical issues in an MSI context. Furthermore, the potential amalgamation of the research scientist and clinician is discussed, bringing innovative bench technologies into the clinical setting.

Collaborative research, networking, and scientific partnerships are crucial in the progression of exciting techniques such as MSI, and this book is a testimonial to the union of likeminded "mass spec" scientists. The authors who have kindly contributed to this protocol series on MSI comprise past and present colleagues and fellow MSI enthusiasts, with many of whom I have enjoyed enlightening scientific discussion whether it be at a conference or simply over a coffee.

Mass spectrometry has enabled me to add the "edge" to my biomedical themed research interests and forge many friendships and collaborations, thus allowing me to connect with scientists throughout Europe, America, and Australia which for me began in my current place of work, the Centre for Mass Spectrometry Imaging Mass Spectrometry Imaging, directed by Professor Malcolm R. Clench within the Biomolecular Sciences Research Centre, Sheffield Hallam University, lead by Professor Nicola Woodroofe

Ultimately the authors herein work directly or in collaboration with mass spectrometry and aim to push the boundaries of biological molecular mapping from atoms to supramolecular complexes.

Sheffield, South Yorkshire, UK *Laura M. Cole*

Contents

Contributors

ZSOLT ABLONCZY • *Ora Inc, Andover, MA, USA*
DAVID M.G. ANDERSON • *Mass Spectrometry Research Center, Department of Biochemistry, Vanderbilt University School of Medicine, Nashville, TN, USA*
ROBERT BRADSHAW • *Centre for Mass Spectrometry Imaging, Biomolecular Sciences Research Centre, Sheffield Hallam University, Sheffield, UK*
ANNE LISA BRUINEN • *Maastricht MultiModal Molecular Imaging (M4I) Institute, Maastricht University, ER Maastricht, The Netherlands*
DAVID J. CALKINS • *The Vanderbilt Eye Institute, Vanderbilt Vision Research Center, Vanderbilt University Medical Center, Nashville, TN, USA*
RICHARD M. CAPRIOLI • *Mass Spectrometry Research Center, Department of Biochemistry, Vanderbilt University School of Medicine, Nashville, TN, USA*
EMMANUELLE CLAUDE • *Waters Corporation, Wilmslow, UK*
MALCOLM R. CLENCH • *Centre for Mass Spectrometry Imaging, Biomolecular Sciences Research Centre, Sheffield Hallam University, Sheffield, UK*
LAURA M. COLE • *The Center for Mass Spectrometry Imaging, Biomolecular Science Research Center, Sheffield Hallam University, Sheffield, South Yorkshire, UK*
ROSALIE K. CROUCH • *Department of Ophthalmology, Storm Eye Institute, Medical University of South Carolina, Charleston, SC, USA*
EVA CUYPERS • *KU Leuven Toxicology and Pharmacology, Leuven, Belgium*
REBECCA E. DAY • *Centre for Mass Spectrometry Imaging, Biomolecular Sciences Research Centre, Sheffield Hallam University, Sheffield, UK*
BENOIT FATOU • *Univ. Lille, Inserm, U1192—Protéomique Réponse Inflammatoire Spectrométrie de Masse—PRISM, Lille, France*
GREGORY L. FISHER • *Physical Electronics, Inc., Chanhassen, Minnesota, United States*
BRYN FLINDERS • *Maastricht Multimodal Molecular Imaging Institute (M41), University of Maastricht, Maastricht, The Netherlands*
ISABELLE FOURNIER • *Univ. Lille, Inserm, U1192—Protéomique Réponse Inflammatoire Spectrométrie de Masse—PRISM, Lille, France*
JULIEN FRANCK • *Univ. Lille, Inserm, U1192—Protéomique Réponse Inflammatoire Spectrométrie de Masse—PRISM, Lille, France*
MANUEL GALLI • *Department of Medicine and Surgery, Proteomics and Metabolomics Unit, University of Milano-Bicocca, Monza, Italy*
PHILIPPA J. HART • *Waters Corporation, Wilmslow, UK*
SARAH HAYWOOD-SMALL • *Biomolecular Sciences Research Centre, Sheffield Hallam University, Sheffield, UK*
RON M.A. HEEREN • *Maastricht Multimodal Molecular Imaging Institute (M41), University of Maastricht, ER Maastricht, The Netherlands*
GÉRARD HOPFGARTNER • *Life Sciences Mass Spectrometry Group, Department of Inorganic and Analytical Chemistry, University of Geneva, Geneva, Switzerland*
EMRYS A. JONES • *Waters Corporation, Wilmslow, UK*

VINCENZO L'IMPERIO • *Department of Medicine and Surgery, Pathology, University of Milano-Bicocca, San Gerardo Hospital, Monza, Italy*
WENDI LAMBERT • *Department of Ophthalmology and Visual Sciences, Vanderbilt University Medical Center, Nashville, TN, USA*
FULVIO MAGNI • *Department of Medicine and Surgery, Proteomics and Metabolomics Unit, University of Milano-Bicocca, Monza, Italy*
NICCOLÒ MOSELE • *Department of Medicine and Surgery, Proteomics and Metabolomics Unit, University of Milano-Bicocca, Monza, Italy*
SARAH NAYLOR • *Department of Health and Wellbeing, Sheffield Hallam University, Sheffield, UK*
FABIO PAGNI • *Department of Medicine and Surgery, Pathology, University of Milano-Bicocca, San Gerardo Hospital, Monza, Italy*
IEVA PALUBECKAITE • *Centre for Mass Spectrometry Imaging, Biomolecular Sciences Research Center, Sheffield Hallam University, Sheffield, UK*
EKTA PATEL • *Centre for Mass Spectrometry Imaging, Biomolecular Sciences Research Centre, Sheffield Hallam University, Sheffield, UK*
TIFFANY PORTA • *Maastricht Multimodal Molecular Imaging Institute (M4I), University of Maastricht, ER Maastricht, The Netherlands; School of Pharmaceutical Sciences, University of Geneva, Geneva, Switzerland*
STEVEN D. PRINGLE • *Waters Corporation, Wilmslow, UK*
JUSAL QUANICO • *Univ. Lille, Inserm, U1192—Protéomique Réponse Inflammatoire Spectrométrie de Masse—PRISM, Lille, France*
MICHEL SALZET • *Univ. Lille, Inserm, U1192—Protéomique Réponse Inflammatoire Spectrométrie de Masse—PRISM, Lille, France*
KEVIN L. SCHEY • *Mass Spectrometry Research Center, Department of Biochemistry, Vanderbilt University School of Medicine, Nashville, TN, USA*
CALLIE SEAMAN • *Centre for Mass Spectrometry Imaging, Biomolecular Sciences Research Centre, Sheffield Hallam University, Sheffield, UK*
ARUL N. SELVAN • *Materials and Engineering Research Institute (MERI), Sheffield Hallam University, Sheffield, UK*
ANDREW SMITH • *Department of Medicine and Surgery, Proteomics and Metabolomics Unit, University of Milano-Bicocca, Monza, Italy*
LYNNE SPACKMAN • *Department of Psychology, Sociology and Politics, Sheffield Hallam University, Sheffield, UK*
PAUL J. TRIM • *South Australian Health and Medical Research Institute (SAHMRI), Adelaide, SA, Australia*
EMMANUEL VARESIO • *School of Pharmaceutical Sciences, University of Geneva, Geneva, Switzerland*
MAXENCE WISZTORSKI • *Univ. Lille, Inserm, U1192—Protéomique Réponse Inflammatoire Spectrométrie de Masse—PRISM, Lille, France*
CHRIS WRIGHT • *Allied Health Sciences, London South Bank University, London, UK*

Chapter 1

"A Future Amalgamation Between the Scientist and the Clinician?"

Sarah Haywood-Small

Abstract

Personalized medicine is gaining momentum and analytical methods such as MS are ideally situated to provide coherent imaging of human disease. The cancer research field is already starting to benefit from the MS imaging applications; non-cancerous diseases will hopefully enjoy the same success. Often, the way forward is to embrace several techniques, which can complement and verify each other. This approach can be seen as less confrontational as everyone can play a part in the development of a new working practice. Stakeholders, professional bodies, and instrument manufacturers will be open to maximizing the patient benefit; investment is more likely given the past performance and reputation of the collaborative highly skilled team.

With this in mind, how close are we to a future amalgamation between the scientist and the clinician? Can we accelerate the integration of innovative bench technologies into the clinical setting and bring state-of-the-art imaging diagnostics to the patient bedside or General Practitioners treatment room?

Key words Collaboration, Translational medicine, Clinical setting, Mass spectrometry

1 Introduction

The key factor that drives almost all research is to inform and improve current practice in the clinical setting. Researchers in the field of mass spectrometry (MS) are not an exception. In other words, research aims to have direct patient benefit within a reasonable timescale. While this can be described as idealistic, on some level, researchers strive hard to achieve this as they increase their expanding repertoire of published journal articles. Specific measurable outcomes of research, such as the presentation of data and ideas at conferences, peer-reviewed journal articles, and the invitation to editorial and scientific advisory roles for example are a daily aspect of our scientific careers.

Translating biomedical science into practical applications can be a slow process, given the need for robust and stringent controls that exist. Current funding and governance models tend to be short-term and nationally focused; at times more support is needed

Laura M. Cole (ed.), *Imaging Mass Spectrometry: Methods and Protocols*, Methods in Molecular Biology, vol. 1618, DOI 10.1007/978-1-4939-7051-3_1, © Springer Science+Business Media LLC 2017

to help design and sharing high-impact resources [1]. Generally, this process can be speeded up in some circumstances, for example in the light of planning for an influenza pandemic [2] or indeed the much-needed resources employed in the recent outbreak of Ebola in West Africa [3]. The speed of this response is mostly attributed to resources, money, and widespread support and above all arguably timely communication between distinctive professionals in their fields. Therein lies a common human trait, the ability to establish and maintain collaborative professional relationships. Such networks exist as a working platform to facilitate the constant review of clinical practice, to improve facilities, and to ensure original contributions that may result in direct clinical benefits.

Everyone who considers themselves to be scientifically trained will have some level of expertise of their field in addition to hard-earned experience with a specific technique. Not only do you need to define your research goals, but you must have a method to achieve your objectives. Expertise in the methodology is paramount, often the importance is overlooked in the initial planning stages—as if you are scientifically trained then surely you have the capability of carry out this procedure? It may require a period of optimizing, but the correct person to actually carry put these experiments needs both the expertise and the tenacity to see the project through to the end. Clinical collaborators may be new to MS as a technique and may need to be convinced in the initial stages—taken together it is easy to understand the hard work that goes into such ventures even before the instrument is switched on. The bottom line is that a vibrant and strong working link is at the heart of translational research and this is only possible by the clinical collaborations established.

2 Establishing Collaborations

Personality types that tend to favor openness and receptivity may naturally fall into conversations at meetings, such discussions lead to professional links. Upon acceptance of publication, the author often receives many invitations themselves to review papers and attend conferences as well. Again, new contacts are made. Often, it is a common theme or research agenda that facilitates collaborations. It is this relationship between the clinician and the researcher that is vital, if the two individuals share a common drive to inform practice that is the first initial step. The main hurdles then become resources, in terms of time and money. MS instrument manufacturers also have a key role in providing the technology that is affordable, robust and that meets all the regulations from overseeing clinical bodies.

We all have the same amount of time in any given working week, and is it an unfortunate working practice that most new collaborations need some time to become established, this time is often competing with the existing workloads. MS as a technology can be unappealing to the newcomer due to the amount of training and sophistication that comes along with a different approach. The researcher then can draw on their technical expertise. In any case, most professionals on either side will willingly admit that this means working outside their comfort zone and maybe their working day, of course these demands and pressures fluctuate over a given period.

If both parties see this as an investment, some degree of optimism will ensue. Many conversations occur, but not all result in long-term collaborations. Often, despite a strong desire to contribute toward a research project, the demands and pressures of other tasks take priority. Therefore, if only for a minimum amount of time, it is a case of making a commitment. There is no guarantee that the paper will be accepted or the grant will be successful, but even the process of trying to initiate the research, useful skills will be learned, and the benefits for both parties will start to materialize.

Money is a contentious issue at the best of times and research depends on strong funding streams. The global economic market experiences many changes, but despite this research will always continue. Grants will always be awarded to some successful applicants, charitable agencies often have a strong influence in funding research that may hold promise for current and future patients. One of the main advantages of collaborative research is exposure to different funding streams and interdisciplinary work. Charitable organizations and some private individuals are strongly motivated ad raise funds to support worthwhile research. In the early stages of a collaboration, it is often the need to first reserve some time and then to identify potential routes of financial support and administrative help if needed. More and more organizations are recognizing the need to develop their staff and provide appropriate support.

Clinical partners can provide a true insight into the current practice, in terms of the areas that are in need of research and development. It may benefit their own career progression as well, in terms of obtaining a higher research degree. In the modern world, collaborative work is based around successful teams and your collaborator may also unlock a new group of interested parties, creating a varied and highly skilled collaborative research team. Good negotiation skills often come into play alongside the ability to hold purposeful conversations, which should eventually lead to the practical implications of project design, planning, and execution. Clinical partners are also in a good position to provide biological samples. Tissue can be archived for research purposes and provided it is collected and used within full ethical consent, research can be highly productive given an accessible library of samples and a highly skilled research scientist.

Ideally, the research will have direct patient benefit and will become established in the clinical setting. Notably, The UK professional doctorate programme enables research to be conducted at doctoral level that is relevant both to professional interests and the organisational context. Bringing cutting edge research to the clinic is possible, to achieve this a certain degree of tenacity is needed within the team, negotiations must be robust and clear to ensure and demonstrate the advantage and gain to patients and best working practices. Negations are not a tradeoff and should be approached fairly and positively for mutual gain. Often, the barriers to this process include time, money, staff resources, conflicting views or external competition, or pressure from internal and external parties. A few examples highlight the strong work ethics of specific individuals, professionals who are united in a common agenda to bring about a real change in practice. Such work is personally rewarding and this behavior should be modeled and cascaded down at every opportunity.

Mentoring schemes should be in place to share good practice, ways in which colleagues can truly learn from their peers. Governing and managerial bodies recognize the importance of developing their staff ad fostering collaborations and so it is worth finding out about the institutional policy and training opportunities. Conferences and organized meetings are an ideal platform to compare regional and cross-disciplinary working practice. The digitalized age means that people do not even have to meet in person, given the popularity of webinars.

Digital pathology is a good example of flexibility within working practice; this intervention allows an internationally renowned pathologist to assess histopathological samples remotely and communicate directly with the operating theatre. Digital slides are used in pathology for education, diagnostic purposes (clinicopathological meetings, consultations, revisions, slide panels, and, increasingly, for upfront clinical diagnostics), and archiving [4]. The improvement in patient outcome and the operational cost effectiveness helped to build ensure this continues to be a good investment to hospital administrators [5].

Conflicting views can also influence the appearance of new and emerging technologies in the clinic. The classic example is the local authority invests in a new technology and trains staff as appropriately, but this is met with some degree of resistance. By admission, most professional colleagues have their own preferred way of working, for example a strong preference for histology in the diagnosis of a particular disease. The individual may have received extensive historic training in this area and may themselves be seen as an expert in this area. The notion to potentially recommend a new imaging technique, for example, may be seen as though the new approach is in some way replacing the current practice, rendering the expertise as redundant. It is likely that everyone will agree with the statement that we need to constantly review and evaluate clinical practice, but the human element is difficult to predict in some circumstances and this may in turn hinder the launch of cutting-edge technology in practice.

Professional bodies will also have an influence in driving change and will play a key part in the emergence of new technologies and working practices. Recommendations should be evidence based and with an ideal goal to improve patient outcome. Cost savings are also highly desirable, efficiency is the key and the collaborative research teams may have the advantage on terms of having expertise at their fingertips.

The last 20 years has seen major breakthroughs in the field of MS. Many advantages related to its ability to image pathological samples and many articles evidence the perceived benefits within the cancer research field. The emergence of MS in the clinical laboratory began initially geared toward newborn screening and steroid analysis. MS has directly improved the practice of medicine and enabled clinicians to provide better patient care. With increasing economic pressures and decreasing laboratory test reimbursement, MS offers new hope and can provide cost-effective solutions [6].

Notably, MALDI-MSI has recently gained considerable attention and has been employed for research and diagnostic purposes, with successful results. It is possible to identify biological markers for multiple proteins in a single MALDI-IMS-MSI experiment. MS images of the distribution of proteins in fresh frozen and formalin-fixed paraffin-embedded tissue samples are established within the cancer field [7]. MS-based proteomics with increased specificity, coverage, and throughput will be pervasive in proteomics investigations in brain cancer research [8]. However, data dimensionality remains a significant challenge accompanied by the statistical methods for information-preserving data reduction. Incorporating relational information may provide a way forward and improve the discriminatory capability of the data [7]. The instrument manufacturers also play a crucial role as they can strive to produce a tool that can be routinely applied in the field of hospital-based clinical pathology [8].

Personalized medicine is gaining momentum and analytical methods such as MS are ideally situated to provide coherent imaging of human disease. The cancer research field is already starting to benefit from the MS imaging applications, non-cancerous diseases will hopefully enjoy the same success. Often, the way forward is to embrace several techniques, which can complement and verify each other. This approach can be seen as less confrontational as everyone can play a part in the development of a new working practice. Stakeholders, professional bodies, and instrument manufacturers will be open to maximizing the patient benefit, investment is more likely given the past performance and reputation of the collaborative highly skilled team. The desire to change and constantly improve is firmly embedded as the cornerstone of medical research. It is for this reason that technologies such as MS will benefit the patients and in fulfilling this goal, it will be rewarding for us and our clinical partners as well.

References

1. Misha A, Schofield PN, Bubela TM (2016) Sustaining large-scale infrastructure to promote pre-competitive biomedical research: lessons from mouse genomics. N Biotechnol 33(2):280–294
2. Barnett DJ, Balicer RD, Lucey DR, Everly GS Jr, Omer SB, Steinhoff MC et al (2005) A systematic analytic approach to pandemic influenza preparedness planning. PLoS Med 2(12):e359. doi:10.1371/journal.pmed.0020359
3. Cohen J (2014) Ebola vaccine: little and late. Science 345(6203):1441–1442. doi:10.1126/science.345.6203.1441
4. Al-Janabi S, Huisman A, Van Diest PJ (2012) Digital pathology: current status and future perspectives. Histopathology 61:1–9. doi:10.1111/j.1365-2559.2011.03814.x
5. Ho J, Ahlers SM, Stratman C, Aridor O, Pantanowitz L, Fine JL, Kuzmishin JA, Montalto MC, Parwani AV (2014) Can digital pathology result in cost savings? A financial projection for digital pathology implementation at a large integrated health care organization. J Pathol Inform 5:33
6. Jannetto PJ, Fitzgerald RL (2016) Effective use of mass spectrometry in the clinical laboratory. Clin Chem 62(1):92–98
7. Cole LM, Mahmoud K, Haywood-Small S, Tozer GM, Smith DP, Clench MR (2013) Recombinant "IMS TAG" proteins – a new method for validating bottom-up matrix-assisted laser desorption/ionisation ion mobility separation mass spectrometry imaging. Rapid Commun Mass Spectrom 27:2355–2362. doi:10.1002/rcm.6693
8. Tian Q, Sangar V, Price ND (2016) Enormous and underutilized potential for brain cancer research. Mol Cell Proteomics 15:362–367

Chapter 2

Fresh Frozen Versus Formalin-Fixed Paraffin Embedded for Mass Spectrometry Imaging

Ekta Patel

Abstract

Matrix-Assisted Laser Desorption Ionization (MALDI) Mass Spectrometry Imaging (MSI) is fast becoming an industry leading technique as a means of investigating the distributions of protein and peptide molecules directly from sections of tissue. Developing protocols for the analysis of FFPE tissue opens up numerous opportunities for novel biomarker discovery, due to the large number of tissue banks containing FFPE biopsy samples. This chapter reports the analysis of both fresh frozen and formalin-fixed paraffin-embedded (FFPE) tissues using such protocols.

Key words MALDI-MSI, Fresh frozen, Formalin-fixed paraffin embedded (FFPE), Peptides, Trypsin

1 Introduction

MALDI-MSI has been successfully used to elucidate the relative abundance and spatial distribution of a number of molecules; ranging from small molecules, peptides, and proteins. The first stage of the MSI workflow is sample collection (Fig. 1). Extra care and detail must be taken from this point through to imaging the sample, as it is crucial that the integrity in spatial localization of the molecules is maintained. It must be emphasized that delocalization and degradation of the analytes may be caused by poor handling and negligence in storage. To avoid such scenarios, it is generally advised that consolidated protocols are followed to ensure the samples are collected in the correct manner. Additionally, the orientation in which the organ is preserved is of great importance to the imaging experiment; disrupting the morphology can lead to ambiguity of results.

Snap-freezing using liquid nitrogen at the point of collection is the most common method employed [1]. Cracking and fragmentation of the tissue may occur as the tissue cools down at different rates. To counteract this, Schwartz et al. (2003) recommend wrapping the tissue loosely in aluminum foil to stabilize the cooling [2].

Laura M. Cole (ed.), *Imaging Mass Spectrometry: Methods and Protocols*, Methods in Molecular Biology, vol. 1618,
DOI 10.1007/978-1-4939-7051-3_2, © Springer Science+Business Media LLC 2017

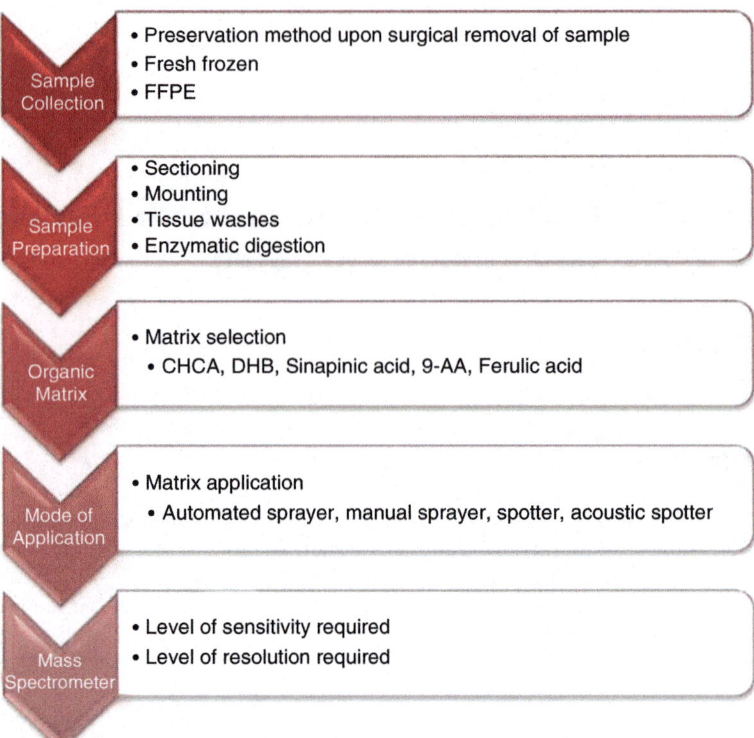

Fig. 1 Basic workflow of an imaging experiment with key focus areas; sample collection, sample preparation, choice of organic matrix , mode of application, and finally choice of mass spectrometer for analyses

On the other hand, Goodwin et al. (2008) mention that wrapping can distort the shape of the tissue, and therefore suggest a method of using ethanol or isopropanol at ≤−70 °C as a solution [3]. We obtained best results when floating a weighing boat containing our tissue onto liquid nitrogen. Fresh frozen tissue may remain frozen at −80 °C without signs of degradation for at least a year.

The alternative to fresh frozen preparations is to formalin-fix paraffin-embed tissue. It was thought that this form of preservation could limit the type of analysis carried out to peptide analysis alone, as the cross-linking in FFPE may prevent the detection of analytes such as metabolites, lipids, and pharmaceutical compounds, however this is not the case [4–6]. Drexler et al. (2007) reported the detection of small molecules in their study which indicates that FFPE may not limit compound analysis [7]. More recently, the Walch group have successfully imaged metabolites from FFPE tissue blocks; their method involves the deparaffinization of FFPE tissue and matrix coating by 9-aminoacridine prior to MALDI-MSI [8].

As the difficulties presented in analyzing FFPE tissues (due to complex cross-linking) are now successfully being overcome, further studies using the numerous samples in tissue banks would be of significant benefit [9]. Analysis would provide an insight into the histology of disease tissues with the potential to discover biomarkers by MALDI-MSI.

2 Materials

All solutions were prepared using ultrapure water (18 MΩ-cm at 25 °C) and analytical grade reagents with a purity of ≥99.5%. Reagents should be disposed of in compliance with health and safety regulations and in accordance with COSHH guidelines.

2.1 Materials for Mass Spectrometry Sample Preparation and Acquisition

1. Dewar of liquid nitrogen.
2. Carboxymethylcellulose (CMC).
3. Cork circles.
4. Cryostat chuck.
5. Surgipath X-tra adhesive precleaned micro slides.
6. Ethyl alcohol, 200 proof, ACS reagent ≥99.5%.
7. Chloroform ≥99.5%.
8. Trypsin Gold, Mass Spectrometry Grade.
9. Ammonium bicarbonate, BioUltra, ≥99.5%.
10. Octyl-α/β-glucoside 10 mM solution (OcGlu).
11. Methanol, CHROMASOLV®, for HPLC, ≥99.9%.
12. Formalin solution, neutral buffered, 10%.
13. Embedding cassettes.
14. Xylene substitute.
15. Embedding medium paraffin wax.
16. Hydrogen peroxide solution (H_2O_2).
17. Trizma base, ≥99.9%.
18. α-Cyano-4-hydroxycinnamic acid (CHCA), ≥98% powder.
19. Aniline, ACS reagent ≥99.5%.
20. Acetonitrile, CHROMASOLV® Plus, for HPLC, ≥99.9%.
21. Phosphorus red, 99%.

2.2 Reagents—Working Composition

1. 2% Carboxymethylcellulose.
2. Ethanol solutions: Dilute 100% ethanol with ultrapure water to 90% and 70% (vol/vol).
3. Ammonium bicarbonate 50 mM: Dissolve 0.2 g ammonium bicarbonate in 50 mL ultrapure water. Check and adjust pH to 8.0. Solution can be stored for 1 month.

4. Trypsin solution 20 μg/mL: Add 1 mL ammonium bicarbonate (pH 8.0) directly to the trypsin 100 μg vial to give a 100 μg/mL stock solution. Agitate to ensure lyophilized trypsin is reconstituted. For the 20 μg/mL working solution, pipette 200 μL of the stock solution into a sterile 1.5 mL snap top Eppendorf and add 800 μL ammonium bicarbonate (pH 8.0).
5. Trypsin solution 20 μg/mL containing detergent 0.5%: Add 5 μL detergent solution (10 mM) to 995 μL trypsin solution 20 μg/mL.
6. Ethanol solutions for paraffin embedding: Dilute 100% ethanol with ultrapure water to 70%, 80%, and 95% (vol/vol).
7. MeOH: H_2O_2 3% solution: Mix together 48.5 mL and 1.5 mL H_2O_2.
8. Tris–HCl buffer (1 L): Completely dissolve 1.21 g Trizma base in 1000 mL of ultrapure water. Adjust pH to 9.0 with 1 M NaOH or HCl as required (solution can be stored in a glass duran for up to 6 months).
9. MALDI matrix solution: Prepare 5 mg/mL CHCA in 50% (vol/vol) acetonitrile and 0.5% (vol/vol) trifluroacetic acid in ultrapure water. Add equimolar amounts of aniline to the CHCA (i.e., 1 mL of 5 mg/mL CHCA solution contained 2.4 μL aniline). Sonicate in a sonic water bath for 5–10 minutes until matrix crystals have fully dissolved. Matrix containing aniline should be prepared fresh on the day of the experiment and not stored.
10. Phosphorus red calibrant: 10 mg/mL in 100% acetone. Vortex for 10–15 seconds. N.B. phosphorus red may not dissolve entirely.

3 Methods

3.1 Cryo-conservation of Tissue

1. Arrange the excised organ/biopsy into the correct orientation onto a regular weighing boat.
2. While wearing appropriate PPE, float the weighing boat containing the tissue on liquid nitrogen (*see* **Note 1**).
3. Take ~50 μL of 2% CMC and attach the frozen tissue to the cork circle, ensuring the tissue is level and in the correct orientation.
4. Place into the −20 °C freezer briefly to ensure that the CMC has frozen.
5. Mount the cork circle with tissue to a cryostat chuck. This can be done with ~50–100 μL 2% CMC solution.

3.2 Sectioning and In Situ Digestion of Fresh Frozen Tissue Sections

1. Fresh frozen rat brain tissue sections should be cut to 10 μm thickness using a Leica CM3050 cryostat chilled to −20 °C.
2. Thaw-mount sections onto Surgipath X-tra adhesive glass slides and store in glass slide holders at −80 °C (*see* **Note 2**).

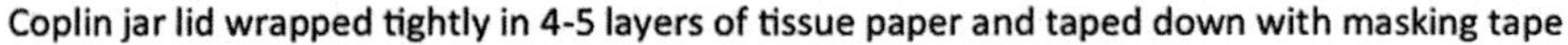

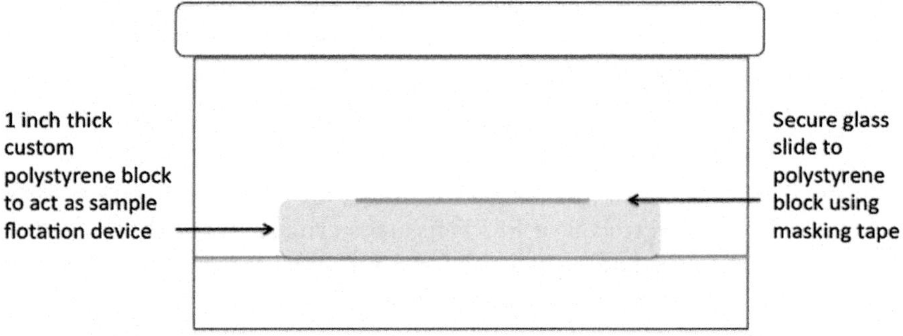

Fig. 2 Humidity chamber setup using a glass coplin jar containing 50:50 MeOH:water and a polystyrene floatation device to hold the sample. Once the lid is placed on the jar, carefully wrap the seal with parafilm to allow the humidity to remain within the chamber

3. Before digestion with trypsin, pretreat tissue sections with wash steps of 70% and 90% ethanol (1 minute each) and chloroform (10 seconds) (*see* **Note 3**).
4. Spray 20 μg/mL trypsin solution containing 0.5% 10 mM OcGlu onto the tissue sections using the SunCollect (SunChrom GmbH, Freidrichsdorf, Germany) automated sprayer (seven layers at a flow rate of 1.5 μL/minute and a nitrogen pressure of 3 bar).
5. Incubate tissue sections in a parafilm-covered coplin jar containing the humidity chamber solution; 50% MeOH (Fig. 2), for 3 hours at 37 °C.

3.3 Tissue Fixation and Embedding

1. Cover the excised tissue with 10% formalin, ensuring the tissue is fully submerged in the fixative. Fix for 24 hours at room temperature.
2. Place the fixed tissues into embedding cassettes, taking care to place the tissue in the correct orientation.
3. Dehydrate the tissue in a series of EtOH concentrations. Place the cassette containing tissue in 70% EtOH for 1 hour, 80% EtOH for 1 hour, 95% EtOH for 1 hour, and finally 100% EtOH for 1 hour.
4. Place the tissue into 100% xylene substitute for 1 hour. Repeat this two more times (therefore totalling 3 hours in xylene substitute).
5. Add liquid paraffin wax to the cassettes and wait for 2 hours (58–60 °C) (*see* **Note 4**). Top up with more paraffin wax if necessary.

3.4 Sectioning and In Situ Digestion of FFPE Tissue Sections

1. Prior to sectioning place the blocks in the −20 °C freezer for 1 hour (*see* **Note 5**). Trim the paraffin blocks on a Leica microtome; sections were cut to 3 μm thickness.

2. Gently place paraffin ribbons (*see* **Note 5**) on a float bath heated to 30 °C. After separating individual sections, mounte onto Surgipath X-tra adhesive micro slides.
3. Air dry sections for 30 minutes and then place into cardboard slide holders. Place holders in a 40 °C oven overnight and store the following day in plastic slide containers.
4. Prior to trypsin digestion, deparaffinization and antigen retrieval of FFPE tissue is crucial. Place the glass slide in 100% xylene substitute for 5 minutes. Repeat using fresh xylene substitute for another 5 minutes.
5. Hydrate the tissue sections in a series of EtOH solutions. Place the glass slide in 100% EtOH for 3 minutes, 95% for 3 minutes, and 70% for 3 minutes.
6. Place the slide into a block to prevent protein degradation. Immerse in ~50 mL MeOH: ultrapure water (3%) for 15 minutes at room temperature.
7. For antigen retrieval, place glass slide in a 50 mL falcon tube and fill with Tris–HCl buffer. Place in a preheated water bath (97 °C) for 30 minutes (*see* **Note 6**).
8. Rinse in ultrapure water.
9. 20 μg/mL trypsin solution containing 0.5% 10 mM OcGlu was sprayed onto the tissue sections using the SunCollect automated sprayer (seven layers at a flow rate of 1.5 μL/minutes and a nitrogen pressure of 3 bar).
10. Tissue sections were incubated in a parafilm-covered coplin jar containing the humidity chamber solution; 50% MeOH (Fig. 2), for 3 hours at 37 °C.

3.5 Matrix Deposition

1. Spray five layers of 5 mg/mL CHCA containing aniline onto the samples previously digested with trypsin, using the SunCollect autosprayer at a flow rate of 1.5 μL/minute and a nitrogen pressure of 2.5 bar (*see* **Note 7**).
2. Ideally, samples should be analyzed on the same day as spraying the matrix. If overnight storage is required due to unforeseen circumstances, keep sample in a sealed petri dish at +4 °C.

3.6 Instrumentation and Data Acquisition

1. Spot 0.5 μL phosphorus red calibrant onto a clean MALDI target plate. Calibrate the SYNAPT™ G2 HDMS mass spectrometer (Waters Corporation, Manchester, UK) using the phosphorus red spot across the 600–2800 Da mass range.
2. Scan the sample sprayed with matrix on a traditional flat-bed scanner and save the image as a .jpeg.
3. Follow the dialog box instructions for a *"new image pattern,"* upload the .jpeg within the high-definition imaging (HDI) software to allow coregistration and image setup (*see* **Note 8**).

4. Load the sample into the instrument, and acquire MALDI-IMS-MS images in positive ion mode from 600 to 2800 at a mass resolution of 10,000 FWHM.
5. Operate the instrument with a 1 kHz Nd:YAG laser and set the laser energy to 220 arbitrary units on the instrument.
6. Perform image acquisition at a spatial resolution of 100 μm × 100 μm for the imaging of tissues.

3.7 Data Analysis

1. Process mass spectrometry imaging data using the HDI software (Waters Corporation, Manchester, UK).
2. It is also possible to use open-source software such as Biomap (Novartis) (*see* **Note 9**).
3. HDI software has a number of features including normalization, color gradients, and the option to create overlays of different ions. N.B. Please see the user manual for an extensive guide.

4 Notes

1. Floating the tissue on liquid nitrogen rather than submerging minimizes the chances of freeze fracturing. This allows the tissue to freeze evenly throughout without damaging. At this stage, the tissue can be covered in parafilm and stored in an air tight container if required at a later date.
2. Thaw mounting: once a section has been cut using the cryostat the tissue section is attached to the glass slide electrostatically; holding the slide close over the section allows it to attach to the glass slide. Rubbing the underside with a gloved finger allows the section to adhere to the slide.
3. Place the 70% and 90% EtOH in a 50 mL falcon tube if one slide is being analyzed at a time. If more than one slide is to be washed at once use a multislide rack in a glass coplin jar. Ensure that the chloroform is always placed in a glass vessel.
4. Melt the paraffin wax pellets at 58–60 °C in a glass beaker.
5. Placing FFPE blocks in the −20 °C freezer for 1 hour before sectioning enables easier sectioning and the production of ribbons. This is a consecutive number of sections (usually 5–10) all attached together. Once the ribbon is placed in the float bath the individual sections can be detached with forceps easily.
6. Carrying out antigen retrieval in a water bath as opposed to a microweave eliminates the increased production of bubbles, which can result in the tissue section detaching from the glass slide.

7. Ensure the Sun Collect has been flushed through with 100% ACN at a flow rate of 2 μL/minute for 15 minutes before matrix is flushed through. This allows any matrix crystals blocking capillaries to be dissolved.
8. The HDI software includes automated image processing. This processing feature can be selected during image setup.
9. Images acquired using the SYNAPT™ can be converted using ImageConverter to be viewed on open-source software such as Biomap (Novartis).

Acknowledgments

The author gratefully acknowledges Dr Simona Francese for her support and guidance, and GlaxoSmithKline for the BBSRC Industrial Case Award funding.

References

1. Goodwin RJA, Nilsson A, Borg D, Langridge-Smith PRR, Harrison DJ, Mackay CL, Iverson SL, Andrén PE (2012) Conductive carbon tape used for support and mounting of both whole animal and fragile heat-treated tissue sections for MALDI MS imaging and quantitation. J Proteomics 75:4912
2. Schwartz SA, Reyzer ML, Caprioli RM (2003) Direct tissue analysis using matrix-assisted laser desorption/ionization mass spectrometry: practical aspects of sample preparation. J Mass Spectrom 38:699
3. Goodwin RJ, Pennington SR, Pitt AR (2008) Protein and peptides in pictures: imaging with MALDI mass spectrometry. Proteomics 8:3785
4. Longuespee R, Casadonte R, Kriegsmann M, Pottier C, Picard de Muller G, Delvenne P, Kriegsmann J, De Pauw E (2016) MALDI mass spectrometry imaging: a cutting-edge tool for fundamental and clinical histopathology. Proteomics Clin Appl 10:701
5. Pietrowska M, Gawin M, Polańska J, Widłak P (2016) Tissue fixed with formalin and processed without paraffin embedding is suitable for imaging of both peptides and lipids by MALDI- IMS. Proteomics 16:1670
6. Angel PM, Caprioli RM (2013) Matrix-assisted laser desorption ionization imaging mass spectrometry: in situ molecular mapping. Biochemistry 52:3818
7. Drexler DM, Garrett TJ, Cantone JL, Diters RW, Mitroka JG, Prieto Conaway MC, Adams SP, Yost RA, Sanders M (2007) Utility of imaging mass spectrometry (IMS) by matrix-assisted laser desorption ionization (MALDI) on an ion trap mass spectrometer in the analysis of drugs and metabolites in biological tissues. J Pharmacol Toxicol Methods 55:279
8. Buck A, Ly A, Balluff B, Sun N, Gorzolka K, Feuchtinger A, Janssen K, Kuppen PJ, Velde VD, Weirich G, Erlmeier F, Langer R, Aubele M, Zitzelsberger H, Aichler M, Walch A (2015) High-resolution MALDI-FT-ICR MS imaging for the analysis of metabolites from formalin-fixed, paraffin-embedded clinical tissue samples. J Pathol 237:123
9. Gorzolka K, Walch A (2014) MALDI mass spectrometry imaging of formalin-fixed paraffin-embedded tissues in clinical research. Histol Histopathol 29:1365

Chapter 3

Imaging MS of Rodent Ocular Tissues and the Optic Nerve

David M.G. Anderson, Wendi Lambert, David J. Calkins, Zsolt Ablonczy, Rosalie K. Crouch, Richard M. Caprioli, and Kevin L. Schey

Abstract

The visual system is comprised of many specialized cell types that are essential for relaying sensory information about an animal's surroundings to the brain. The cells present in ocular tissue are notoriously delicate, making it particularly challenging to section thin slices of unfixed tissue. Maintaining the morphology of the native tissue is crucial for accurate observations by either conventional staining techniques or in this instance matrix-assisted laser desorption ionization (MALDI IMS) or imaging using mass spectrometry. As vision loss is a significantly debilitating condition, studying molecular mechanisms involved in the process of vision loss is a critically important area of research.

Key words Matrix-assisted laser desorption ionization, Imaging mass spectrometry, Rodent ocular tissue, Optic nerve, Ocular pathology

1 Introduction

Matrix-assisted laser desorption ionization imaging mass spectrometry (MALDI IMS) or imaging using mass spectrometry methodologies has seen significant advances in recent years, with improvements in both instrumentation and sample preparation techniques allowing for high-quality, high spatial resolution data to be acquired from single cells and small anatomical structures [1–3]. This high spatial resolution molecular information allows for the mapping of molecular spatial distributions occurring in fine structures, such as the cells that comprise ocular tissues [4–8]. The highly organized cellular structure of the neural retina is made up of multiple cell types of varying dimensions (~10–80 μm) in close proximity. The main function of these cells is to relay information gathered by the light-sensitive photoreceptors to the brain through the optic nerve. Disruption of cellular processes in this region of the body can result in a variety of blinding diseases. Due to the small size and varying densities of the tissues present in the eye globe, special considerations need to be made to prevent

Laura M. Cole (ed.), *Imaging Mass Spectrometry: Methods and Protocols*, Methods in Molecular Biology, vol. 1618, DOI 10.1007/978-1-4939-7051-3_3, © Springer Science+Business Media LLC 2017

compression and stretching of these regions during the sectioning process. Obtaining sections from fresh frozen tissues, while preserving the morphology, is crucial so that they are suitable for high spatial resolution MALDI IMS, in order to reveal the native distribution of molecular information.

Ice crystal formation during freezing has long been known to affect the resulting morphology of fresh frozen tissues. Rapid freezing prevents ice crystal growth and subsequent damage to the tissue. Liquid nitrogen provides the lowest temperature to freeze tissue in a laboratory environment, although immersing tissue directly into liquid nitrogen results in the outside of the tissue freezing much faster than the central regions. This causes expansion of the central regions and resulting in a pressure differential in the tissue and freeze fracturing of the tissues during sectioning. Rapid freezing of ocular tissues using liquid nitrogen indirectly (liquid nitrogen cooled environment) preserves the morphology of the native tissue and prevents freeze fracturing of the tissue.

The rodent lens occupies a large proportion of the ocular globe and therefore it presents several problems for the analysis of the neural retina or cornea from rodent ocular tissue. The density of the lens affects section quality, notably around the central region of the ocular globe, with its density being greatest at the central core region. As a result, stretching during sectioning of the surrounding neural retina and cornea occurs and gives poor quality sections. Another problem posed by the lens tissue is the vast quantities of protein (up to 300 mg/mL) present in the fiber cells of the tissue. This high concentration of protein prevents successful washing of the tissue, as the proteins are difficult to completely remove and often contaminate the surrounding tissue regions. This contamination may result in ion suppression or high background signals, producing misleading results. Removal of the rodent ocular lens without disruption to the morphology of the surrounding tissue from a freshly dissected or unfrozen tissue is challenging, and this is addressed in the following paragraphs.

Analysis of molecular information from the optic nerve of rodent tissue using high spatial resolution imaging mass spectrometry also requires specific considerations based on the size of this tissue. Many animal models of ocular disease have been developed in mice. The diameter of a mouse optic nerve is approximately 400 μm and its length can vary depending on the dissection. Cutting sections of optic nerve tissue, with small dimensions from distal to proximal at the region of the optic nerve, requires alignment in the same axis along the length of this tissue. Alignment is important because, an un-straightened nerve is free to travel in and out of the sectioning plane, requiring analysis of many sections to gather information from all regions distal to proximal.

One way to improve section quality and reproducibility is embedding of the tissue; however, conventional embedding compounds such as optimal cutting temperature media (OCT) or paraffin wax are not suitable for the MALDI ionization process without special sample preparation techniques. Carboxyoxymethyl cellulose (CMC) has been demonstrated on a number of tissue types to help in the sectioning process [4, 9]. Orientation of the sample in an embedding medium so that sections can be obtained along the optic nerve tissue from proximal to distal in the same section also poses a challenge.

In this chapter, we will discuss sample preparation methods for ocular tissue from rodents to help preserve the morphology of the neural retina and prevent contamination from proteins present in the ocular lens. An example is provided that demonstrates alignment of the optic nerve tissue allowing for longitudinal sections along the optic nerve to the optic nerve head with the neuronal tissue passing into the neural retina.

2 Materials

2.1 Sample Preparation

1. Carboxymethyl cellulose sodium salt.
2. Hypodermic needle, gauge 19–21.
3. Cryostat.
4. Large weighing boat $5^1/_2'' \times 7/8''$ H.
5. 100 mM ammonium acetate.
6. 2,5-Dihydroxybenzoic acid (DHB).
7. Nikon Eclipse 90i microscope.
8. Craftman 8 in. drill press.

2.2 Data Acquisition and Processing

1. IMS data was acquired using a Bruker Ultraflextreme II TOF/TOF mass spectrometer.
2. Data acquisition and image processing was performed using Fleximaging 4.1.

3 Methods

3.1 CMC Embedding Methods for Rodent Ocular Tissues

Embedding has long been the method of choice to improve quality of sections for pathological interpretation or immunohistochemistry techniques, although many of these techniques are not suitable for analysis by mass spectrometry without special sample preparation methodologies. OCT is a compound embedding medium comprised of polymeric molecules such as polyethylene glycol (PEG). This medium provides a dense resilient solid material once frozen, and provides sections with limited stretching or compression of the tissue as it passes across the blade. Due to the chemical composition of OCT ions, suppression of tissue-derived signals is commonly observed without prior washing of the section. CMC is also polymeric in nature, but unlike OCT, does not suppress MALDI signal and was first demonstrated by the analysis of whole-body animal sectioning [9]. CMC also provides a dense medium with resilient properties once frozen.

3.2 Preparation of Carboxymethyl Cellulose Solution

Dissolving CMC powder in deionized water can take many hours, depending on the concentration. The concentration should be optimized based on the sample type, with higher concentrated solutions (up to 5%) better suited for denser tissues (*see* **Note 1**). Practical concentrations for rodent and human ocular tissue are 1–2.7%

1. Weigh out the desired amount of CMC and place in a suitable bottle along with a magnetic stir bar, and shake the solution to ensure the powder becomes immersed.
2. Place the bottle onto a hot plate at 70 °C and initiate the magnetic stirrer at a moderate speed.
3. Leave the solution on the hot plate until CMC power is completely dissolved, checking periodically to ensure the stir bar is still cycling and shaking to dislodge large clumps of undissolved material.
4. Once dissolved, allow to cool to room temperature. Degas the solution prior to use, using a sonication bath (*see* **Note 2**).
5. The solution should be stored at 4 °C to prevent microbial growth and sonicated to degas before each use.

3.3 Embedding of Eye Tissue for Neuronal Retina Analysis Using CMC

Because the neural retina is a very delicate tissue, rapid freezing of the tissue is essential to retain the morphology in its native form. Since immersing room or body temperature tissue directly into liquid nitrogen results in freeze fracturing of tissue, freezing in a liquid nitrogen cooled environment provides the most optimal freezing times. CMC is also affected by freezing time, affecting the resulting structural integrity of the support material; better results are observed with faster freezing times.

1. Care should be taken to gently enucleate the ocular globe from the sacrificed animal including as much of the optic nerve if required. Freeze enucleated eye immediately on a large plastic weighing boat floating on top of liquid nitrogen in a desired orientation.
2. Once the tissue is frozen throughout (freezing time depends on sample size) CMC solution can then be conservatively applied to the sample, ensuring it does not defrost the tissue, using a 1 mL or Pasteur pipette, coating the tissue surface first on one side, the weigh boat can be removed briefly from the liquid nitrogen to better observe the sample.

3. Once the CMC from the first application has frozen more can be added and frozen if needed. The sample can be inverted using precooled forceps before adding more CMC to the opposite side to encase the tissue.

3.4 CMC Embedding for Optic Nerve Analysis with Alignment Using a Hypodermic Needle

1. Once enucleated the eye can be mounted onto a medium gauge (19–21G) needle, slightly piercing the globe around the region of the optic nerve head. The remaining optic nerve tissue should be straightened along the needle, marking the Luer lock region in the place aligned with the tissue as seen in Fig. 1a.
2. The tissue should then be frozen by immersing the base of the needle into liquid nitrogen without immersing the actual tissue, while holding with long forceps as seen in Fig. 1b, after 1–2 min, the pupil will take on an opaque appearance once the lens has frozen.
3. The tissue on the end of the needle can then be quickly immersed into the CMC solution contained in a weighing boat with suitable depth and the sample can now be immersed in liquid nitrogen to encase the tissue. This process can be repeated a number of times until sufficient material is surrounding the sample.
4. The sample should be handled with care, as any twisting of the sample or the needle could result in the mark on the Luer lock used to indicate the tissue region becoming displaced.

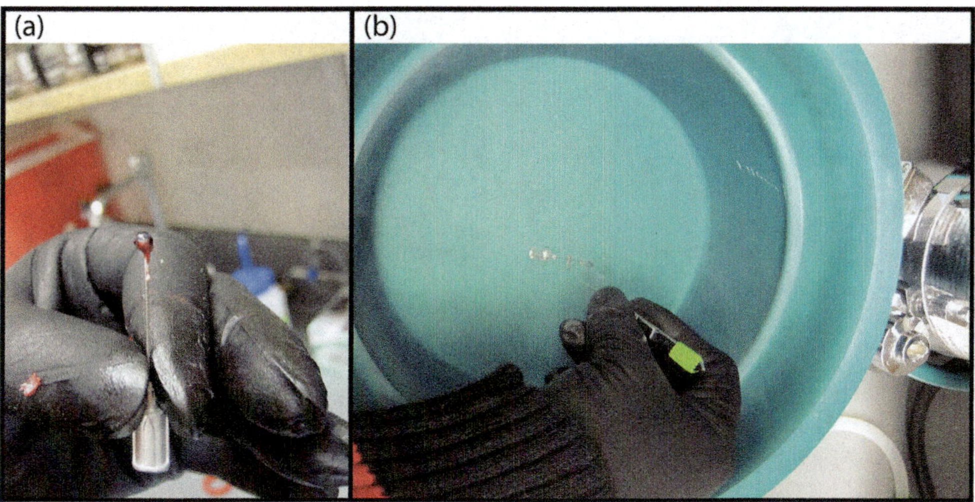

Fig. 1 Image displaying (**a**) alignment of mouse optic nerve tissue along hypodermic needle after dissection, (**b**) displaying initial freezing process where the lower portion of the hypodermic needle is immersed into liquid nitrogen using long handled forceps before immersing in CMC solution

3.5 Cryostat Sectioning of Ocular Tissue

3.5.1 Obtaining Sections for Neural Retina and Cornea Analysis

1. Samples should be mounted onto the cryostat chuck using OCT in the desired orientation; a flat surface can be shaved onto the mounting CMC surface using a razor blade to ensure correct orientation and good adherence.
2. The angle of the stage should be adjusted to allow for a less acute passage of the tissue section over the blade as seen in Fig. 2, as acute angles cause the surrounding CMC to fragment and lose structural integrity.
3. CMC should be trimmed using larger section increments of 30–50 μm until the tissue becomes visible.
4. Before section procurement, the blade should be replaced as trimming can cause blade dulling that will diminish section quality. Once the tissue is visible, excess CMC can now be removed from around the sample using a precooled razor blade if necessary.
5. Once a desired region for sectioning is reached, the section can be manipulated into position on the target surface using a fine-tipped paint brush on the surrounding CMC material, preventing contamination or disruption of the tissue.
6. Flat sections thaw mount better than wrinkled sections as wrinkled sections thaw at different rates across the tissue. This can cause tissues to stretch and fold, affecting the morphology and resulting in artifacts in the MALDI IMS data. One way to prevent wrinkled and stretched sections is to remove the lens, *see* Subheading 3.6.

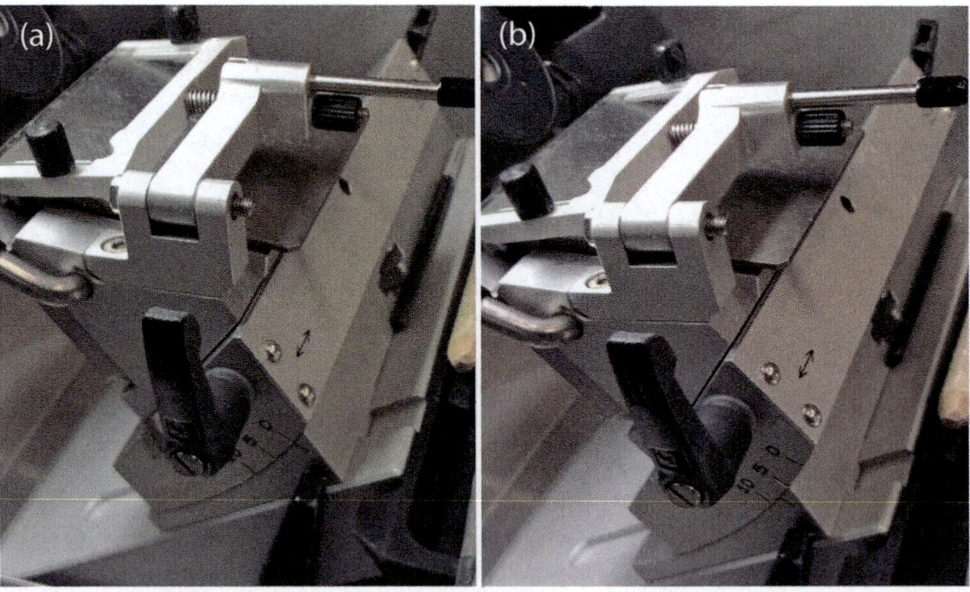

Fig. 2 (**a**) Image showing less acute angle on a Leica CM 3050 S sample stage. (**b**) Image showing more acute angle on a Leica CM 3050 S sample stage

7. Once thaw mounted onto a target plate or slide, the section should be air dried immediately, while providing heat from a gloved hand before returning the sample to the cryostat or to a vacuum desiccator.

3.5.2 Obtaining Sections from Aligned Optic Nerves

1. Attach the sample to the sample chuck, with the mark on the Luer lock indicating the position of the tissue on the needle facing downward toward the sample chuck. The sample should be frozen to the sample chuck using minimal OCT, keeping the needle as parallel to the chuck as possible as seen in Fig. 3.
2. Once attached the chuck the sample can be placed in the sample chuck holder. The sample can be advanced toward the blade using the manual control adjustment on the cryostat to bring the sample closer to the blade and sample stage. Make adjustments to the sample holder until the needle is as visually parallel to the blade as possible.
3. The sample can now be trimmed to allow the needle to get closer to the blade to make adjustment more accurate, taking care not to let the luer lock collide with the blade, Fig. 4. Once satisfied with the adjustment the needle can be pulled from the embedded sample using a gentle twisting motion. Continue with trimming until the impression left behind by the needle is visible on the block face as seen in Fig. 5, initially the widest region of the impression indicates closest region to the blade, fine adjustments can be made to ensure the sample alignment is optimal. The cryostat blade should be replaced at this point as trimming of the

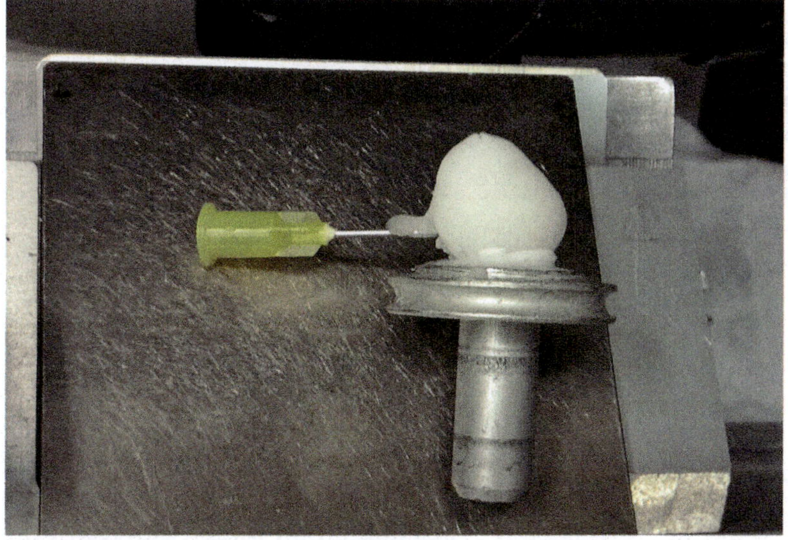

Fig. 3 Image of CMC embedded mouse eye attached to a cryostat chuck with the mark on the luer lock of the needle oriented toward the chuck and mounted with the needle parallel to the chuck surface

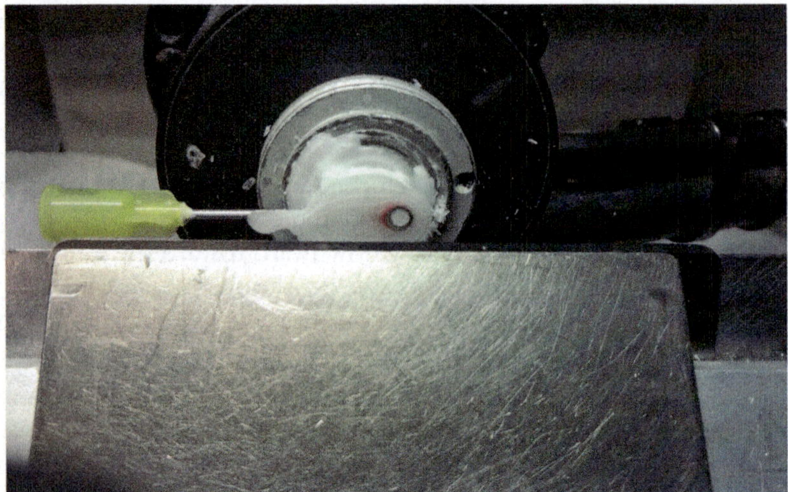

Fig. 4 Image of partially sectioned CMC embedded mouse eye with the hypodermic needle still in place, so it can be aligned parallel to the cryostat blade

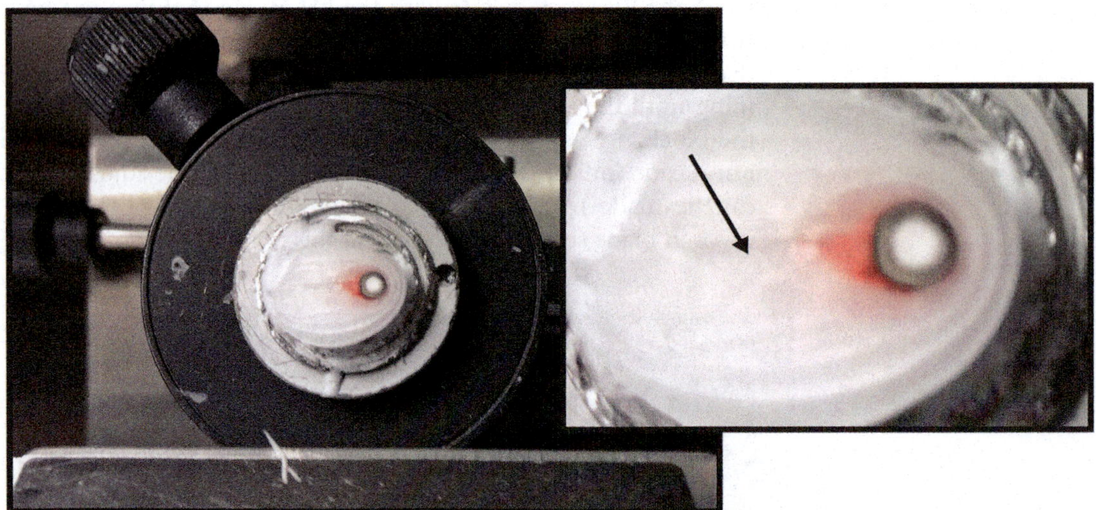

Fig. 5 Image of CMC embedded mouse eye once the hypodermic needle removed and partially trimmed to the point where the needle impression is visible, indicated by the arrow in the zoomed inlay

CMC results in premature dulling of the blade. The position of the anti-roll bar can also be optimized at this point. Section procurement can begin after the impression is no longer visible.

4. The surrounding CMC material around the tissue should have sufficient structural integrity to manipulate the section into position on the sample surface (if the material has poor structural integrity, *see* **Note 3**).
5. The CMC should prevent tissue curling although wrinkles may still be observed in larger sections. These wrinkles can affect

section quality, wrinkles turn into folds during the thaw mounting process. Adjusting the anti-roll bar position may help to prevent wrinkles and gently flatting the tissue before thaw mounting will prevent folds.

3.6 Removal of the Lens from Frozen Tissue

The lens is a very large, dense region of the rodent eye tissue in comparison to the neural retina or cornea. Sectioning of the mid region of the eye to obtain imaging mass spectrometry data from the neural retina or cornea, free from sectioning artifacts such as wrinkles or folds, can be challenging. The lens from rodent eyes can be gently removed using a fine-tipped paint brush, although the success rate of thaw mounting such a section and retaining good morphology is low. One method to overcome these issues is to physically remove the lens from an embedded, and partially sectioned, eye using a drill press and dry ice, ensuring the tissue remains frozen. This will improve both section quality and reproducibility.

1. The embedded sample should first be trimmed on a cryostat to expose the majority of the lens on the block face of the embedded tissue. A suitable sized drill bit should be precooled on dry ice before mounting into the drill press (Fig. 6a).
2. The embedded eye should be accurately placed in the path of the drill bit which can be checked before turning on the drill. The majority of the lens tissue can then be drilled out, taking care not to damage the surrounding neural retina or cornea, *see* Fig. 6a.

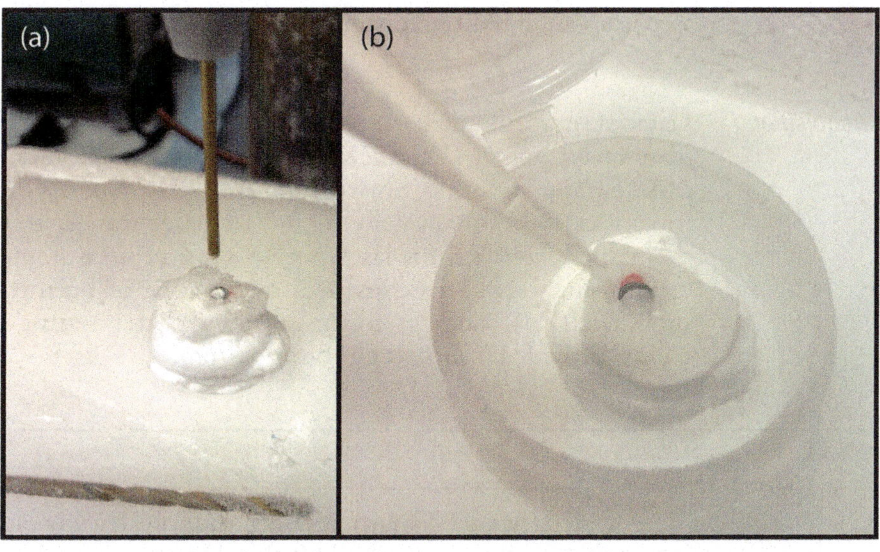

Fig. 6 (**a**) Image of CMC embedded mouse eye placed on a block of dry ice below a 7/64″ drill bit mounted into a pillar drill used to remove lens tissue from a partially sectioned mouse eye. (**b**) Image of mouse eye with the lens tissue removed and the void backfilled with CMC solution

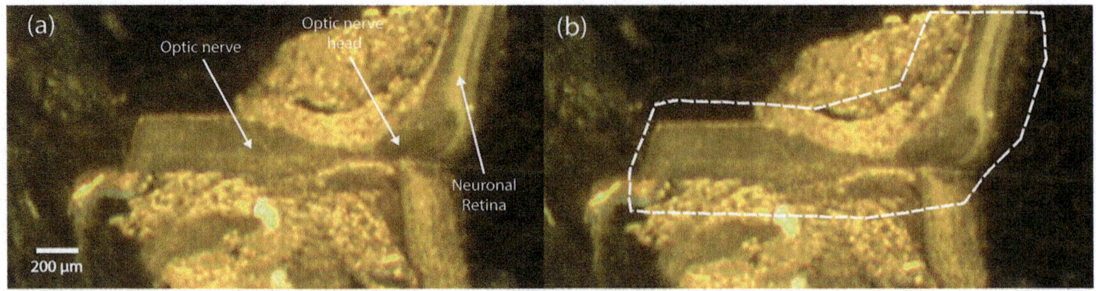

Fig. 7 (**a**) Optical image of a section mounted onto a gold-coated MALDI sample plate taken from mouse eye indicating the hypodermic needle aligned optic nerve, optic nerve head, and neural retina regions. (**b**) Optical image of the same section indicating region where data were acquired

3. The hole should then be back filled with CMC solution using a 1 mL pipette while keeping the tissue on dry ice, Fig. 6b. Trim the protruding CMC from the hole before continuing sectioning.

3.7 Tissue Washing and Matrix Application for Lipid Analysis

In the example provided for this method, the optical image in Fig. 7 displays a section taken from a mouse optic nerve, optic nerve head, and neural retina. The section was thaw mounted onto a gold-coated MALDI target plate and air dried immediately prior to vacuum desiccation. The section was then dehydrated in a vacuum desiccator overnight before washing three times for 30 s each with 100 mM ammonium acetate. The sample was then dried again for another 2 h in a vacuum desiccator. 2,5-Dihydroxybenzoic acid (DHB) was deposited by sublimation onto the samples using a custom sublimation apparatus for 20 min at 54 mTorr pressure at 120 °C.

3.8 Data Acquisition

Data were acquired in positive ion mode using a MALDI-TOF mass spectrometer (UltrafleXtreme II, Bruker Daltonics, Billerica, MA) equipped with a Smartbeam II 1 kHz Nd:YAG frequency tripled to 355 nm wavelength, and the laser was set to the minimum spot size with 8% laser power. Optical measurements were taken using a Nikon Eclipse 90i microscope of burn patterns in sublimed DHB surfaces to ensure laser power setting and shot counts provided a sub 10 µm burn pattern.

4 Notes

1. A variety of CMC solution concentrations have been reported in the past from 1 to 5%. This should be optimized on a tissue-to-tissue basis with higher concentrations better suited to denser tissues. One disadvantage to using higher concentrations of CMC is premature dulling of the blade.

2. Liquid water and CMC solutions contain a small amount of air that forms bubbles upon freezing. These bubbles affect the quality of the frozen CMC substrate. The gas dissolved in the liquid can be removed by degassing prior to use. This can be either performed with a vacuum degasser or via sonication.
3. Poor structural integrity or striations observed in the embedding medium can be a result of poor anti roll bar position or angle of the sample stage having been incorrect. Anti-roll bar placement, too close to the sample results in "chattering" thicker sections are also more prone to this effect.

5 Results

The optical image Fig. 7 displays a section taken from a mouse optic nerve and retinal tissue where the nerve was aligned along a hypodermic needle thaw mounted onto a gold-coated MALDI target plate. The transition of the optic nerve into the ocular globe at the optic nerve head indicated by the arrows is clearly visible. The image in Fig. 7b displays the region where data were acquired and includes the optic nerve region with some of the surrounding connective tissue, the optic nerve head, and a small portion of the neural retinal layers. The data in Fig. 8 display lipid distributions from multiple anatomical regions of a mouse optic nerve and neural retina. The signal at *m/z* 592.5 in Fig. 8a observed in red in the central overlay, previously identified as A2E [4], defines the retinal pigment epithelium in the neural retina, above the point where the optic nerve meets the ocular globe at the optic nerve head. The optic nerve and optic nerve head regions are clearly defined by a signal at *m/z* 760.6 in Fig. 8b (white in central overlay) previously identified as PC(16:0_18:1) [5]. The distribution displayed by this signal is highly abundant in the optic nerve and traverses the optic nerve head and passes through into the inner and outer plexiform, inner nuclear and ganglion cell layers of the neural retina. A signal observed at *m/z* 734.5 in Fig. 8c displays high abundance in the outer nuclear layer of the neuronal retina, with lower intensity signal for this lipid observed in the inner plexiform and nuclear layers of the neural retina. The signal observed for an unidentified ion at *m/z* 762.6 displays high abundance in the inner plexiform and nuclear layers of the neural retina, displayed in green in the central overlay, this signal spans up to the optic nerve head region of the tissue. This distribution observed in green in the central overlay suggests an association with the neuronal tissues of the neural retina, but not the optic nerve. The signal at *m/z* 834.7, displayed as yellow in the central overlay, is present in high abundance in the photoreceptor layer of the tissue, with lower intensity spanning into the inner layers of the neural retina. The spatial resolution of

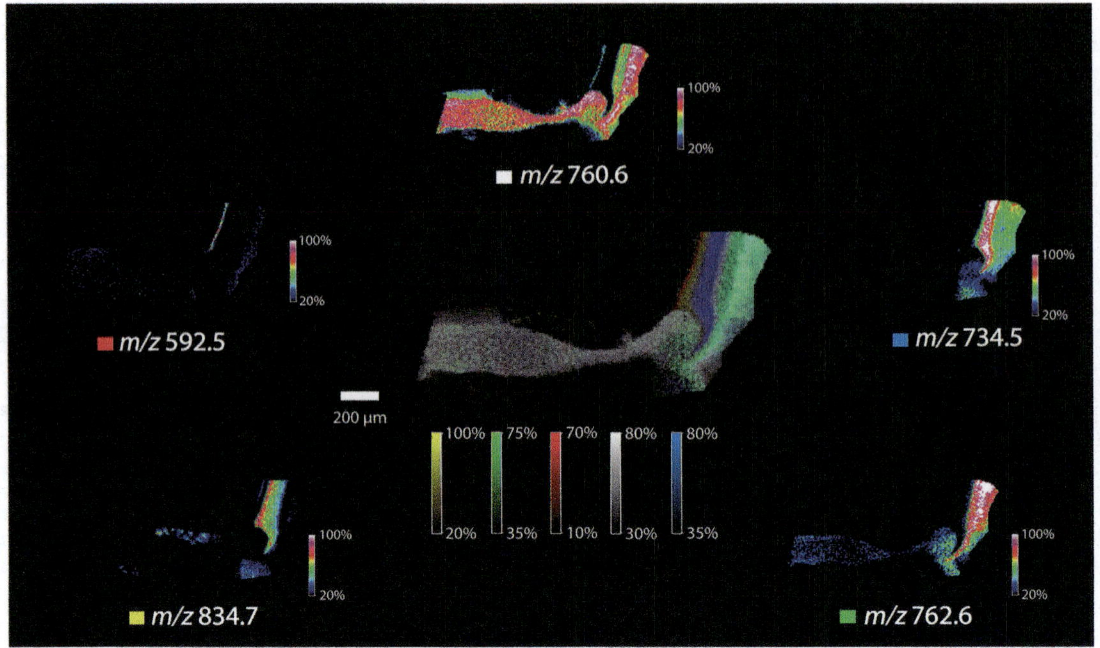

Fig. 8 MALDI IMS data acquired in positive ion mode of mouse optic nerve and neural retina. Ion images display the distributions of (**a**) A2E, *m/z* 592.5, (**b**) PC(16:0_18:1), *m/z* 760.6, (**c**) unidentified ion, *m/z* 734.6, (**d**) unidentified ion, *m/z* 762.6, and (**e**) unidentified ion, *m/z* 834.7. The central image is an overlay of the individual ion maps shown

10 µm, in this example, provides high signal intensity from small anatomical regions of mouse ocular tissue, thus providing MALDI images that display clear definition of multiple cell layers in the neuronal retina and clear definition of the optic nerve and optic nerve head.

6 Conclusions

Embedding ocular tissues in frozen degassed CMC greatly improves section quality and reproducibility from section to section preserving the morphology of the native tissue, which improves the quality of MALDI IMS data obtained. This embedding method allows for successful and reproducible sections from the neural retina and cornea regions of the tissue. Further manipulation by removing the ocular lens and refilling the space further improves section quality from more central regions of the ocular globe and prevents proteins from lens tissue contaminating the neural retina or cornea regions during washing protocols. Aligning the optic nerve along a hypodermic needle allows data to be acquired along the full length of the tissue in one section allowing for comparisons of molecular abundances in distal and proximal regions, including the optic nerve head, to be compared in the same data set. The methods described here can be easily adapted to be suitable for other tissues analyzed by MALDI IMS, which are problematic in similar areas, such as section quality or alignment of small features.

References

1. Zavalin A et al (2012) Direct imaging of single cells and tissue at sub-cellular spatial resolution using transmission geometry MALDI MS. J Mass Spectrom 47(11):1473–1481
2. Grove KJ et al (2014) Diabetic nephropathy induces alterations in the glomerular and tubule lipid profiles. J Lipid Res 55(7):1375–1385
3. Schober Y et al (2012) Single cell matrix-assisted laser desorption/ionization mass spectrometry imaging. Anal Chem 84(15):6293–6297
4. Anderson DM et al (2014) High resolution MALDI imaging mass spectrometry of retinal tissue lipids. J Am Soc Mass Spectrom 25(8): 1394–1403
5. Anderson DM et al (2013) High-resolution matrix-assisted laser desorption ionization-imaging mass spectrometry of lipids in rodent optic nerve tissue. Mol Vis 19:581–592
6. Anderson DM et al (2015) High spatial resolution imaging mass spectrometry of human optic nerve lipids and proteins. J Am Soc Mass Spectrom 26(6):940–947
7. Zemski Berry KA et al (2014) Spatial organization of lipids in the human retina and optic nerve by MALDI imaging mass spectrometry. J Lipid Res 55(3):504–515
8. Bowrey HE et al (2016) Imaging mass spectrometry of the visual system: advancing the molecular understanding of retina degenerations. Proteomics Clin Appl 10(4): 391–402
9. Stoeckli M, Staab D, Schweitzer A (2007) Compound and metabolite distribution measured by MALDI mass spectrometric imaging in whole-body tissue sections. Int J Mass Spectrom 260(2–3):195–202

Chapter 4

MALDI-MSI of Lipids in Human Skin

Philippa J. Hart and Malcolm R. Clench

Abstract

Matrix-assisted laser desorption ionization (MALDI) mass spectrometry (MS) is now a well-established technique for imaging analysis of sectioned biological tissues. One of the growing areas of interest is in the analysis of skin. MALDI-MSI can provide a wealth of information from within sections of skin. This includes information on the distribution of pharmaceuticals following topical treatments, through to the examination of the composition of different skin layers and studies of proteomic, lipidomic, and metabolomic responses to disease, wounds, and external stimuli. Here, we describe the handling procedures, preparatory treatment, and mass spectrometry setup required for the MALDI MSI analysis of lipids within human skin samples.

Key words MALDI, MS, MSI, Imaging, Skin, Lipids

1 Introduction

Over the last decade the application of MALDI-MS imaging (MALDI-MSI) for skin analysis has seen a significant increase. In addition to the skin penetration studies that are well documented to include MALDI MSI analysis [1], there is also much evidence to suggest the studies of skin lipids (SLs) that have been carried out [2, 3] are of value due to the importance of lipids in biological systems and their involvement in disease pathogenesis. A large number of SLs and their associated metabolites are localized within the stratum corneum and epidermal regions of the skin. These lipid species are directly affected as a result of the proteomic behavior within skin [4] and as such, are also involved in disease pathogenesis [5]. It has been established that sphingosine (Sph), sphingosine-1-phosphate (S1P), ceramide (Cer), and ceramide-1-phosphate (C1P) have pivotal roles in apoptosis and other cell signaling pathways [6]. In particular, C1P and S1P are claimed to have a role in the inflammation pathway and the arachidonic acid cascade [7, 8].

As an example, MALDI-MS has also been previously used to analyze skin biopsies taken from patients with Fabry's disease. In this

Laura M. Cole (ed.), *Imaging Mass Spectrometry: Methods and Protocols*, Methods in Molecular Biology, vol. 1618,
DOI 10.1007/978-1-4939-7051-3_4,

case, a role for glycosphingolipids in disease pathogenesis was observed as they are found to accumulate in the skin of patients affected with the disease [3]. Other MALDI-MSI studies of human skin include those into the mechanisms of wound healing, for example, subsequent to burns [9]. The many matrices that can be employed in the use of MALDI-MSI make it ideally suited for the analysis of multiple classes of lipids and as such an ever-valuable tool.

2 Materials

In every case, solutions must be prepared using ultrapure, deionized water and analytical grade reagents used, of purity 99.8% or above. Unless otherwise stated, all chemical reagents were stored at room temperature in a suitable container in a solvent spill tray. All reagents should be disposed of in a manner compliant with health and safety regulations and handled according to all COSHH guidelines.

2.1 Tissue Embedding

1. 2% Carboxymethylcellulose in water (*see* **Notes 1** and **2**).
2. A large Blister pack (vacuum formed, plastic tray), or other suitable mold.
3. Liquid nitrogen, in a suitable, insulated container.
4. A large, plastic weighing boat.

2.2 Tissue Sectioning

1. Polylysine-coated (*see* **Note 3**) glass slides (If a conductive surface is required indium tin oxide-coated glass slides should be used).
2. 2% Carboxymethylcellulose in water (*see* **Notes 1** and **2**).
3. Cryostat with new, clean blade fitted.
4. Methanol

2.3 Washing Solutions

1. Ultrapure (deionized) water, cooled at 4° prior to use.
2. Glass coplin jar.
3. Vacuum desiccator.

2.4 MALDI Matrix

1. Ultrapure (deionized) water.
2. Methanol.
3. Ethanol.
4. >98% Trifluoroacetic acid (TFA) should be added to deionized water to give a solution of 0.2% TFA.
5. Sonicator.
6. Automated spraying device.
7. α-Cyano-4-hydroxycinnamic acid (CHCA)/ANI solution: Dissolve 5 mg of CHCA in 70:30, ethanol: 0.2% TFA solution and vortex (sonication can be used if any CHCA remains

undissolved). Next, add an equimolar amount of aniline, as described by Calvano et al. [10], to the matrix solution and sonicate for 5–10 min (*see* **Notes 4–6**).

3 Methods

3.1 Tissue Embedding

1. All preparation of human skin should be performed in a category 2 laminar flow hood in order to protect both the user and the fidelity of the sample.
2. Blocks of skin can be cut to size using biopsy punches, or a scalpel where a biopsy punch is not applicable.
3. Skin tissue blocks were placed into individual blister packs filled with carboxymethyl cellulose (CMC), ensuring that the tissue blocks were fully immersed.
4. The blister packs were then snap-frozen by placing the samples in a weighing boat, floating on liquid nitrogen, not directly into the liquid nitrogen. This allows for easy handling and helps to avoid freezer damage.
5. The embedded samples were then removed from the blister packs and placed into labeled petri dishes or centrifuge tubes, which were sealed with parafilm.
6. All frozen samples were stored in sample boxes, clearly labeled as containing human tissue, in the −80 °C freezer.

All human tissue remains and waste materials that had been in contact in any way with human tissue were placed inside a biohazard bag in the −80 °C to await disposal by incineration. Any non-disposable implements were disinfected in bleach.

3.2 Tissue Sectioning

1. Set the cryostat temperature to −20 °C.
2. Set the cutting thickness of the cryostat to 10–12 μm.
3. Use a small amount of the prepared CMC gel to mount a flat surface of the sample onto the specimen stage, ensuring that the sample is in an appropriate position to give the desired orientation of cutting (*see* **Note 7**). Mount the specimen stage onto the cryostat.
4. Cut through the embedding material until an even cross section of the sample is reached. If there is a large amount of embedding material present, this can be removed by using a scalpel or by increasing the cutting thickness (e.g., to 30 μm) (*see* **Note 8**).
5. Once a section is made, this should be placed onto the glass slide and the "freeze-thaw" method used to mount the section onto the slide. This involves placing a gloved finger on the underside of the glass slide to warm the surface, thus adhering the tissue to the slide.

6. Subsequent to sectioning, slides that are not immediately analyzed should be placed in a sealed container and stored at −80 °C until used.

3.3 Tissue Washing

Prior to coating with matrix it may sometimes be preferable to wash tissue to remove salts, particularly if the tissue is ex-vivo or has been stored in buffer solution prior to sectioning.

1. Fill a glass coplin jar with cooled deionized water (*see* **Note 9**).
2. Immerse the slide with the mounted section(s) into the water for 20–60 s and repeat if necessary, taking care to avoid creating turbulence in the water [11, 12].
3. Subsequent to immersion, tap off excess water and desiccate the sample until dry (e.g., for approximately 5–15 min depending on the environment).

3.4 Matrix Coating

1. The spraying device, in this case, the Suncollect (SunChrom) should first be flushed for a minimum of 15 min with the solvent of the matrix solution, in this case, ethanol.
2. The matrix solution should then be flushed through the sprayer for long enough to go through the lines, e.g., 5–10 min.
3. Set the gas pressure to 2 bar.
4. Check that the sprayer is depositing a smooth, homogenous stream of matrix prior to setting up the device for the tissue section.
5. Teach the software the location of the tissue section; in the case of the suncollect device this involves the recording of the coordinates from the top left and bottom right of the section(s) for spraying.
6. Set the matrix spraying as follows:

   In medium speed mode: 10 layers of matrix were applied using the continual spraying mode at medium speed, with the spraying needle at a distance of 2 cm from the tissue surface. The first two layers were sprayed at a flow rate of 10 μL/min, layers 3–8 at 15 μL/min and the final two layers at 20 μL/min.
7. Start the spray sequence and monitor the quality of deposition for at least the first pass (*see* **Notes 10** and **11**).
8. Subsequent to coating the tissue section the spraying device should again be flushed with solvent to prevent any crystallization in the lines and blockages (*see* **Notes 10–12**).

3.5 MALDI Data Acquisition/ Setup

1. Set acquisition mode to positive or negative mode (If lipids investigated ionize in negative mode, *see* **Note 13**).
2. The mass range for acquisition was set from 50 to 1200.

3. Scan the slide to be analyzed before coregistering and setting up the area for acquisition (*see* **Note 14**).
4. Choose an appropriate value for spatial resolution or pixel size (*see* **Note 15**) and processing parameters (where applicable), prior to commencing the imaging run.

3.6 Software Processing/Image Visualization

A wide variety of software is available for processing and visualization. Some of the freely available imaging tools include, but are not limited to, MALDIvision, MSIreader, Biomap, Datacube explorer. Each MALDI-MS manufacturer also has their own dedicated software for analyzing imaging datasets. In this case, where ion mobility is applied on Synapt instrumentation, the Waters, High Definition Imaging (HDI) software should be used for visualization of the ion mobility dimension in relation to imaging.

1. Choose a suitable color gradient to show the relevant features in your sample.
2. Choose an appropriate form of normalization where required (*see* **Note 16**).

4 Notes

1. Carboxymethylcellulose (CMC) can be used at between 1 and 5% concentrations. In this case, 2% was found to be adequate but increasing the percentage will alter the rigidity of the embedding material.
2. When dissolving the CMC in water it is beneficial to use a magnetic flea for assistance in dissolving the material. This should be done well in advance of sample receipt as it can take some time to dissolve.
3. Alternatives to polylysine-coated glass slides can be used but an adherent coating or positively charged surface is recommended to promote adherence of the tissue on the slide and prevent loss of tissue during washing procedures [1].
4. When using MALDI matrix containing aniline, this should be prepared fresh on a daily basis as aniline is light sensitive and degrades rapidly. This should also be prepared in an amber vial or eppendorf, if these are not available however, a transparent container can be wrapped in aluminum foil to protect the solution from light.
5. As an example, for a 5 mg/mL solution of CHCA a volume of 2.4 μL of aniline was added.

6. If performing MS/MS mode analysis some lipids tend not to fragment well when present as [M+H]+. To improve fragmentation lithium salt can be added to the matrix. For implementation, 100 nM of LiCl [14] was dissolved in 70% EtOH, 0.2% TFA, with α-CHCA being added to the solution to give a concentration of between 5 and 10 mg/mL.
7. When positioning the specimen stage ready for cutting with the cryostat, take care of the orientation. Where a cross section through the skin layers (stratum corneum, epidermis, dermis, etc.) is required, sectioning was found to be easier with the sample in a vertical orientation. This helps to prevent curling or flaking of the different layers and improves the capture of thinner layers.
8. When sectioning be certain to clean the stage and roll plate on a regular basis with methanol to ensure no tissue has adhered and to prevent contamination.
9. In addition, or, as an alternative to water, ammonium acetate or ammonium formate solutions may be used for removal of salts prior to analysis [13].
10. When spray coating a tissue section the deposition should be monitored to ensure that the deposition lays in a fine spray, wetting the tissue enough for matrix:analyte cocrystallization, but not so much as to cause either streaking/stripey effects in the coating or delocalization.
11. If the matrix spray is not smooth and is sputtering, this may be a sign that there is some blockage in the lines. In which case, the lines should be flushed with solvent once more and any parts where there may be potential dead volume should be checked for matrix crystals and cleaned.
12. Sublimation may also be applicable in cases of high spatial resolution imaging [15].
13. For acquisitions in negative ionization mode, alternative matrices are required. Solutions of either dihydroxybenzoic acid (DHB), 1,5- Diaminonapthalene (DAN) [15], or 9-aminoacridine [16] are among the useful matrices in these cases.
14. If you are unsure of the coregistration it is beneficial to set a larger area for the acquisition to avoid cutting out any of the sample.
15. When choosing the spatial resolution for an image it is important to consider that when acquiring at higher spatial resolution, sensitivity will reduce due to the smaller number of ions being desorbed from the smaller acquisition area. It can also be a good idea to perform an acquisition at lower spatial resolution to check whether any ions of interest are detected before moving to higher spatial resolution imaging experiments.
16. With regards to normalization, a paper by Fonville et al. [17] discusses this issue in further depth, comparing different methods of normalizing MALDI MSI, such as normalization against

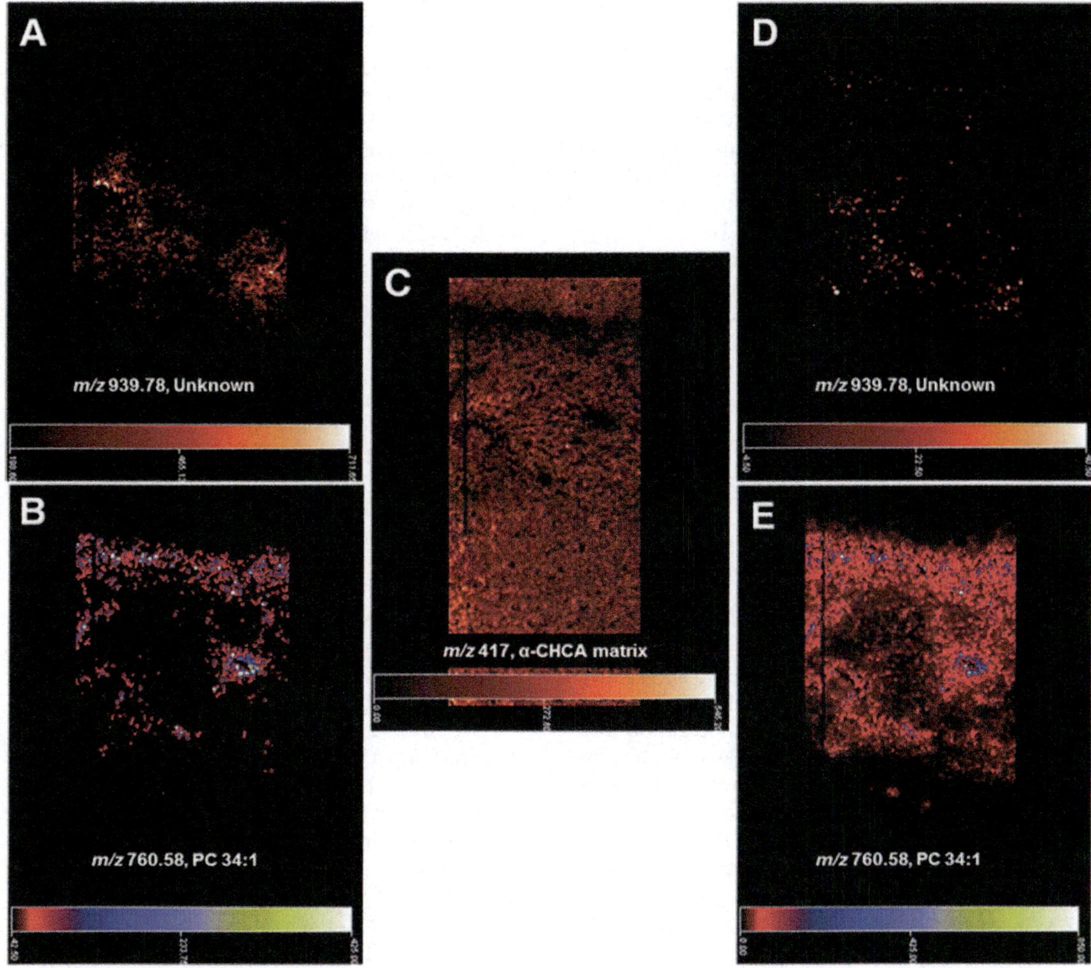

Fig. 1 MALDI-MS images of a cross section of untreated human skin (with the stratum corneum at the top of each image), acquired at a spatial resolution of 30 μm × 30 μm. Images labeled (**b**) (m/z 760) and (**e**) (m/z 939) have been normalized against the matrix peak, m/z 417 (**c**), and the images labeled (**a**) (m/z 760) and (**d**) (m/z 939) have not been normalized [18]

matrix TIC, against a species endogenous to the tissue, against the TIC of all data and to the TIC of a reference molecule/ external standard. Figure 1 shows the effects of choosing a matrix ion where the cocrystallization has varied across features within a section of skin. This highlights the importance of choosing an algorithm, peak or peak list to normalize against which has a homogenous distribution across the sample.

Acknowledgments

This work was supported by the Biomolecular Sciences Research Centre of Sheffield Hallam University.

References

1. Enthaler B, Pruns JK, Wessel S, Rapp C, Fischer M, Wittern KP (2012) Improved sample preparation for MALDI–MSI of endogenous compounds in skin tissue sections and mapping of exogenous active compounds subsequent to ex-vivo skin penetration. Anal Bioanal Chem 402(3):1159–1167
2. Sahle FF, Gebre-Mariam T, Dobner B, Wohlrab J, Neubert RHH (2015) Skin diseases associated with the depletion of stratum corneum lipids and stratum corneum lipid substitution therapy. Skin Pharmacol Physiol 28:42–55
3. Roy S, Touboul D, Brunelle A, Germain DP, Prognon P, Laprévote O, Chaminade P (2006) Imaging mass spectrometry: a new tool for the analysis of skin biopsy. Application in Fabry's disease. Ann Pharm Fr 64(5):328–334
4. Doering T, Hollerans WM, Potratz A, Vielhaber G, Elias PM, Suzuki K, Sandhoff K (1999) Sphingolipid activator proteins are required for epidermal permeability barrier formation. J Biol Chem 274(16):11038–11045
5. Fujiwaki T, Yamaguchi S, Sukegawa K, Taketomi T (2002) Application of delayed extraction matrix-assisted laser desorption ionization time-of-flight mass spectrometry for analysis of sphingolipids in skin fibroblasts from sphingolipidosis patients. Brain Dev 24:170–173
6. Bartke N, Hannun YA (2009) Bioactive sphingolipids: metabolism and function. J Lipid Res 50:591–596
7. Chalfant CE, Spiegel S (2005) Sphingosine 1-phosphate and ceramide 1-phosphate: expanding roles in cell signalling. J Cell Sci 118:4605–4612
8. Pettus BJ, Bielawska A, Subramanian P, Wijesinghe DS, Maceyka M, Leslie CC, Evans JH, Freiberg J, Roddy P, Hannun YA, Chalfant CE (2004) Ceramide 1-phosphate is a direct activator of cytosolic phospholipase A2. J Biol Chem 279:11320–11326
9. Taverna D, Pollins AC, Sindona G, Caprioli RM, Nanney LB (2016) Imaging mass spectrometry for accessing molecular changes during burn wound healing. Wound Repair Regen 24(5):775–785. [epub ahead of print]
10. Calvano CD, Carulli S (2009) Aniline/α-cyano-4-hydroxycinnamic acid is a highly versatile ionic liquid for matrix-assisted laser desorption/ionization mass spectrometry. Rapid Commun Mass Spectrom 23(11):1659–1668
11. Hart PJ, Francese S, Claude E, Woodroofe MN, Clench MR (2011) MALDI-MS imaging of lipids in ex vivo human skin. Anal Bioanal Chem 401:115–125
12. Mitchell CA, Long H, Donaldson M, Francese S, Clench MR (2015) Lipid changes within the epidermis of living skin equivalents observed across a time-course by MALDI-MS imaging and profiling. Lipids Health Dis 14(84):1–12
13. Angel PM, Spraggins JM, Baldwin HS, Caprioli R (2012) Enhanced sensitivity for high spatial resolution lipid analysis by negative ion mode matrix assisted laser desorption ionization imaging mass spectrometry. Anal Chem 84:1557–1564
14. Jackson SN, Wang HJ, Woods AS (2005) In situ structural characterisation of phosphatidylcholines in brain tissue using MALDI-MS/MS. J Am Soc Mass Spectrom 16:2052–2056
15. Thomas A, Charbonneau JL, Fournaise E, Chaurand P (2012) Sublimation of new matrix candidates for high spatial resolution imaging mass spectrometry of lipids: enhanced information in both positive and negative polarities after 1, 5-Diaminonapthalene deposition. Anal Chem 84(4):2048–2054
16. Cerruti CD, Benabdellah F, Laprevote O, Touboul D, Brunelle A (2012) MALDI imaging and structural analysis of rat brain lipid negative ions with 9-aminoacridine matrix. Anal Chem 84:2164–2171
17. Fonville JM, Carter C, Cloarec O, Nicholson JK, Lindon JC, Bunch J, Holmes E (2012) Robust data processing and normalisation strategy for MALDI mass spectrometric imaging. Anal Chem 84:1310–1319
18. Hart PJ (2012) MALDI-MS investigation of skin and its response to irritants and sensitisers. PhD Thesis, Sheffield Hallam University, Sheffield

Chapter 5

MALDI-MSI Analysis of Cytological Smears: The Study of Thyroid Cancer

Niccolò Mosele, Andrew Smith, Manuel Galli, Fabio Pagni, and Fulvio Magni

Abstract

Fine needle aspiration (FNA) biopsies are the current gold-standard for the preoperative evaluation of thyroid nodules. However, a significant number of them (15–30%) are unable to be affirmatively diagnosed and are given an "indeterminate for malignancy" final report, meaning that the malignant nature of the thyroid nodule remains unknown and the recommended therapeutic approach is total thyroidectomy. Furthermore, cytomorphological evaluation of biopsies taken post-surgery indicates that approximately 80% of nodules within this group of patients are in fact benign, and the total thyroidectomy unwarranted. Therefore, the identification of new possible diagnostic targets that can assist in the preoperative diagnosis of thyroid tumors and reduce the number of unnecessary thyroidectomies is imperative.

Matrix-Assisted Laser Desorption/Ionization (MALDI)—Mass Spectrometry Imaging (MSI) has the ability to provide very precise and localized information regarding protein expression in cytological specimens. This enables the detection of cell subpopulations based on their different protein profiles, even within regions that are indistinguishable at the microscopic level, and the feasibility of this approach to investigate *FNA* specimens has already been highlighted in a number of studies. Here, an overview about the sample preparation procedure for the MALDI-MSI analysis of ex vivo FNA biopsies is provided, highlighting how molecular imaging can be combined with traditional histology to generate protein signatures of the different thyroid lesions, and, ultimately, build classification models that can be potentially used to classify benign and malignant thyroid nodules from a molecular standpoint.

Key words Thyroid, Cancer, MALDI-imaging, Proteomics, Cytology

1 Introduction

Thyroid cancer represents the most frequent endocrine malignancy worldwide and its incidence has raised significantly in recent decades, increasing by 8.8% in the United States within the last 65 years [1, 2]. Contributing to this malignancy are the two cell strains that are resident in the thyroid gland: follicular cells, which result in the more common malignancies (papillary thyroid carcinoma (PTC) and follicular thyroid carcinoma (FTC)), and parafollicular C cells, which result in a rarer form of malignancy (medullary

Laura M. Cole (ed.), *Imaging Mass Spectrometry: Methods and Protocols*, Methods in Molecular Biology, vol. 1618,
DOI 10.1007/978-1-4939-7051-3_5,

Table 1
The current comprehension of benign and malignant thyroid nodules

Benign thyroid nodules	Malignant thyroid nodules
Follicular adenoma (FA)	Conventional variant papillary thyroid carcinoma (cvPTC)
Hyperplastic (Hp) Hashimoto's thyroiditis (HT)	Follicular variant papillary thyroid carcinoma (fvPTC)
Uncertain malignant potential (UMP)	Mixed variant papillary thyroid carcinoma (mvPTC)

thyroid carcinoma (MTC)) [3]. Furthermore, many benign thyroid lesions, such as follicular thyroid adenoma (FTA) and Hashimoto thyroiditis (HT), can also affect the thyroid and may represent a more challenging diagnostic aspect for clinicians [4, 5] (Table 1), especially in the preoperative phase. The recommended treatment for many thyroid tumors is total thyroidectomy, and thus the preoperative distinction between the benign and malignant conditions is crucial to avoid inappropriate surgical removal and resultant lifelong hormone replacement therapy.

Fine needle aspiration (FNA) biopsies are the current gold-standard for the preoperative evaluation of thyroid nodules [6]. However, a significant number of them (15–30%) are unable to be affirmatively diagnosed and are given an "indeterminate for malignancy" final report, meaning that the malignant nature of the thyroid nodule remains unknown and the recommended therapeutic approach is total thyroidectomy. Furthermore, cytomorphological evaluation of biopsies taken post-surgery indicates that approximately 80% of nodules within this group of patients are in fact benign, and the performed thyroidectomy was consequently the incorrect therapeutic approach. Therefore, the identification of new diagnostic targets that can assist in the preoperative diagnosis of thyroid tumors, and ultimately reduce the number of unnecessary thyroid removals, is considered to be imperative.

A number of proteomic approaches have been applied for the purpose of detecting new markers that can assist in the diagnostic characterization of thyroid nodules, employing both preoperative FNA biopsies and surgical specimens [7]. In particular, MALDI-MSI has the ability to provide spatially resolved information regarding protein expression in cytological specimens [6]. This enables the characterization of cell subpopulations based on their molecular profiles, even within regions that are indistinguishable at the microscopic level [8], and the feasibility of this approach to investigate ex vivo FNA specimens has already been highlighted [9, 10].

Mainini et al. performed a preliminary study on various types of nodules, including cvPTC, fvPTC, hyperplastic nodules, and Uncertain Malignant Potential (UMP) tumors, and attempted to differentiate these lesions by virtually microdissecting mass spectra from very discrete regions of cell populations associated with the malignant conditions and subsequently performing hierarchical cluster analysis (HCA) and principal component analysis (PCA) [9]. Complementary results were then obtained by Pagni et al. in their feasibility study, applying MALDI-MSI on a further set of nodules, including hyperplastic nodules, Hurtle cell-type FTA, MTC, and cvPTC [10]. The results showed that those nodules with similar biological behavior were represented by similar protein profiles, indicating that MALDI-MSI could potentially be employed to successfully classify these various thyroid lesions from a proteomic standpoint.

Of greater diagnostic relevance, Pagni et al. then attempted to distinguish benign lesions from malignant [11]. In this latest study, ex vivo fine needle aspirations from benign thyroid nodules and papillary thyroid carcinomas were analyzed by MALDI-MSI. Employing the protein profiles obtained from the aforementioned patients as a reference, MALDI-MSI was able to correctly assign, in blind, ten additional FNAB specimens to a malignant or benign class. This data suggested that an approach based on MALDI-MSI has the potential to play a significant role in the diagnosis of thyroid lesions and provide further insights into the development of tumorigenic activity.

Here, an overview on the sample preparation procedure for the MALDI-MSI analysis of ex vivo FNA biopsies is provided (Fig. 1). Although the workflow is similar to what would be required for the analysis of fresh-frozen tissue, there are specific points to consider given that the samples are cytological and can be prone to a number of interfering molecules and structures. Furthermore, given that the most biologically relevant information is present within only a small number of cells within the sample, the collaboration with the pathologist is of utmost importance. It is through the ability of MALDI-MSI to provide specific protein profiles of discrete cell populations, and the close collaboration with the pathologist, that the proteomic information from these small cellular aggregates can be exploited to provide an insight into the malignant nature of the thyroid nodule undergoing diagnosis.

2 Materials

1. 25 G fine needles.
2. Diff-Quick staining solution.
3. Standard microscope glass slides.

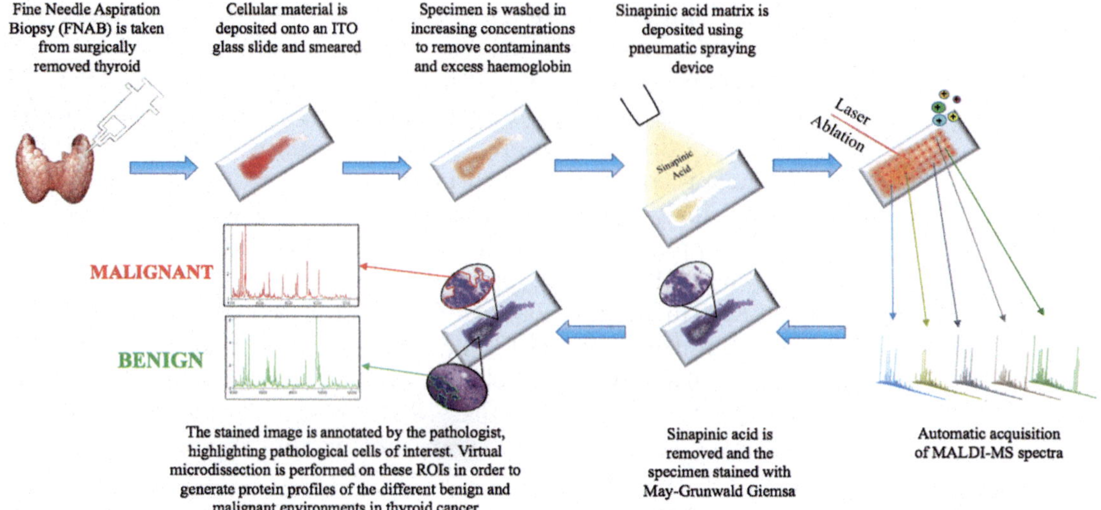

Fig. 1 MALDI-MSI workflow for the analysis of cytological smears in the study of thyroid cancer

4. Indium Tin Oxide-coated glass slides (Bruker Daltonik GmbH, Bremen, Germany), or other MALDI targets (instrument compatible).
5. Ethanol Solutions (A, B, C, D: 70, 90, 95, 100%).
6. SA matrix solution: 10 mg/mL in 60/40 ACN:H_2O w/0.2% TFA.
7. Protein Calibration Standard: Angiotensin II, Angiotensin I, Substance P, Bombesin, ACTH clip 1–17, ACTH clip 18–39, Somatostatin 28 (Bruker Daltonik GmBH, Bremen, Germany).
8. Methanol.
9. Hematoxylin and Eosin.
10. May-Grunwald Giemsa solution (for post-analysis staining of the cytological specimen).

2.1 Instrumentation

1. Standard desktop digital scanner.
2. Scanscope CS2 digital scanner (Aperio, USA).
3. *iMatrixSpray* (for matrix deposition) (Tardo Gmbh, Subingen, Switzerland).
4. MALDI Mass Spectrometer: UltrafleXtreme (Bruker Daltonik GmBH, Bremen, Germany).

2.2 Software

1. *FlexControl* 3.4 (Bruker Daltonik GmBH, Bremen, Germany)—for setting the parameters for spectral acquisition.
2. *FlexImaging* 3.0 (Bruker Daltonik GmBH, Bremen, Germany)—for image acquisition and visualization.
3. *SCiLS Lab 2014* (http://scils.de/; Bremen, Germany)—for extensive data elaboration.

3 Methods

3.1 Sample Procurement and Stocking

1. Cytological samples are obtained from targeted nodules of thyroidectomy specimens, using fine needles (25 G), within 15 min after surgery. This procedure enables the procurement of cytological samples that contain a large number of thyroid cells, without artifacts and with minimal waste of thyroid tissue, to be obtained. Briefly, for each target nodule, different passes are performed in order to ensure that the nature of the nodule is accurately represented. These preliminary passes are stained with Diff-Quick in order to evaluate the adequacy of the cellular material. Once confirmed, the cellular material is drawn from the desired region of the nodule by the fine needle and deposited onto a conductive indium tin oxide (ITO) glass slide. Using a clean microscope glass slide, the droplet containing cellular material is then smeared along the surface of the ITO glass slide.
2. Most commonly, specimens from both the nodule and the benign region of the thyroid, in a location opposite to the nodule, are taken. Given the variability associated with the sampling procedure, this enables more appropriate comparisons between malignant and benign cells.
3. The proteome is then stabilized in a vacuum desiccator for 30 min.
4. The ITO glasses are then packaged and stored in a freezer at −80 °C until the day of MALDI-MSI analysis.

3.2 Sample Preparation for MALDI-MSI

1. Following removal from the freezer, the ITO glass slides onto which the cytological specimens are smeared are dried via desiccation for 30–60 min (based upon the amount of material present). Vacuum dehydration is performed in order to remove excessive moisture from the surface of the specimen that may occur during the freezing and thawing processes, thus reducing the possibility of analyte delocalization. It is during this phase that the composition of the sample can be evaluated. If the color of the sample is particularly red, indicating the presence of hemoglobin, the specimen may require more extensive washing in order to remove this strong ion suppressant.
2. The tissue sections are then placed sequentially into Petri dishes containing increasing concentrations of Ethanol (A, B, C) and gently agitated for 30s during each wash. In instances where the specimen contains high amounts of hemoglobin, or there is the presence of colloidal material, additional washes in Ethanol Solution C may be required in combination with more aggressive agitation (*see* **Notes 1** and **2**). Excess Ethanol is then quickly removed from the slide by gently wiping with tissue paper.

3. Given the translucence of many of the specimens, the smear may be difficult to see in the scanned image; thus, a solvent-resistant marker can be used to highlight the border of the smear on the nonconductive side of the ITO glass.
4. Teaching points are then positioned on the corners of the slide/target prior to scanning. Scanning of the smear is performed prior to matrix deposition in order to obtain a clear image of the specimen which is then used to link the image acquisition software (*FlexImaging*) with the mass spectrometer's software (*FlexControl*) that controls the geometry of the target. The resolution of the scanned image should be selected based upon the spatial resolution of the planned MALDI-MSI analysis (2400 dpi is usually sufficient for 30 μm spatial resolution MALDI-MSI acquisitions).
5. The SA matrix solution is deposited onto the tissue using the *iMatrixSpray* automated spraying device. Given that a balance between spatial localization and sensitivity is required for this application, example spraying parameters are as follows; Spray height: 60 mm, Speed: 180 mm/s, Line distance: 1 mm, Density: 4 μL/cm^2, Break: 30 s and Number of cycles: 10[11]. However, if instrumentation that is capable of obtaining MALDI-MS Images at spatial resolutions of less than 30 μm is being used, and such spatial resolutions are desired by the user, the spray "height," "density," and "break" may need to be adjusted accordingly.
6. Matrix is removed from the edges of the slide using Methanol and tissue paper in order for the conductivity of the slide to be completely maintained. Matrix is also removed from a small area near to the tissue in order to add protein calibrants.
7. A solution containing a 1:1 ratio between SA and the protein calibrant is prepared and 0.8 μL of this calibrant is spotted onto a clean area of the ITO slide/target.

3.3 MALDI-MSI Analysis

1. The previously positioned teaching points are used for teaching the instrument and to set the *X* and *Y*-axes. The measurement region is then set using *FlexImaging* software and information related to the geometry is sent to the instrument using *FlexControl*. Other parameters, such as the rastering, are configured in *FlexImaging* and selected based upon the dimension of the cells to be analyzed and the matrix crystal dimensions. The selected rastering for cytological smears should be considered prior to the analysis, with a compromise between spatial resolution and acquisition time being made. Considering the mean dimensions of a smear, a raster of 80 μm is recommended in order to obtain sufficient proteomic information from the cellular aggregates while ensuring that the acquisition

time does not exceed 12 h (*see* **Note 3**). For each raster position, a mass spectrum is acquired by accumulating between 300 and 500 shots. The number of shots is dependent upon the amount of matrix on the sample and its homogeneity, which can vary due to the inherent differences in the morphology of each specimen.

2. Following calibration using the protein standard, the selected section is analyzed using a Bruker UltrafleXtreme mass spectrometer equipped with a Smartbeam laser (Nd:YAG, 355 nm wavelength, 1–2 kHz frequency) and operating in positive linear mode in the mass range of 3–15 kDa.

3.4 Cytological and Histological Evaluation

1. Following MALDI-MSI analysis, the matrix can be removed from the smear by washing sequentially in increasing concentrations of Ethanol (A, B, D) and gently agitating for 30 s during each wash. Excess Ethanol is then quickly removed from the slide by wiping gently with tissue paper.
2. The smear is then stained with May-Grunwald Giemsa or Hematoxylin and Eosin.
3. The slides are then converted to digital format using the ScanScope CS digital scanner in order to directly correlate the molecular information with the particular morphological features.
4. Regions of Interest (ROIs) are selected by the pathologist, favoring areas with a high density of homogeneous and well-preserved cells and excluding areas with artifacts.

3.5 Data Elaboration

1. The acquired data is first visualized in *FlexImaging* for preliminary evaluation and preprocessing. The digital image of the stained slide is coregistered with the molecular image.
2. ROIs are drawn around the pathologically significant cells previously highlighted by the pathologist. Virtual microdissection can be performed in order to extract spectra only from these regions.
3. The data from the desired specimens, including the ROIs, is then be imported into *SCiLS Lab 2014* for more intensive processing and data elaboration.
4. A series of preprocessing steps are performed on the loaded spectra: baseline subtraction (TopHat algorithm) and normalization (total ion current algorithm). A series of further steps are performed in order to generate an average (avg.) spectrum representative of the whole measurement region, of benignity or malignancy, and the different subclasses of thyroid lesions (m/z 3000–15,000): peak picking (orthogonal matching pursuit algorithm), peak alignment (to align the detected ions with peak maxima), and spatial denoizing. The minimal interval width is set at ±4 Da.

5. Principal Component Analysis (PCA) can also be performed to reduce the high complexity of the data and to monitor spectral similarities between disease subclasses. Then, Receiver Operative Characteristic (ROC) analysis is performed to discriminate between the different thyroid lesions and their malignant nature, with an AUC of >0.75 being required, as an additional criterion to the $p < 0.05$ of the intensity variation, for a peak to be considered statistically significant.
6. In order to minimize the effect of signals, such as hemoglobin that are predominant in the regions of the smear that do not contain cells of interest (e.g., Colloidal areas), that may hamper the potential to detect discriminatory ions, the data analysis can be performed by using only the ROIs. Within *SCiLS Lab*, these ROIs can be first grouped together to represent one individual patient and then grouped further in order to represent a particular thyroid lesion.
7. For the classification process, the patients are split into two groups; one for the training phase, most commonly including patients with a clear cytological report (benign and malignant), and another for the test phase that is most commonly represented by patients with an indefinite cytological report but with a clear diagnosis after histological evaluation after thyroidectomy. On the training dataset, using the *k-Nearest Neighbor* (*k-NN*) classifier, feature selection (the maximum number of features is set at 20) is performed in order to find signals (*m/z*) that are able to correctly classify benign and malignant lesions and to discard the redundant and invariant features. For this purpose, the average spectrum for each individual benign and malignant patients, which was built within *SCiLS Lab*, are employed (from *m/z* 3000 to 15,000). The algorithm iteratively selects a subset of features (*m/z*) and, for each particular feature subset selected, evaluates their classification capability, using a tenfold cross-validation on the K-NN model. The features with the best performances (*Average Probability Assigned to the Correct Class;* value from 0 to 100) are selected in order to evaluate the model with the testing dataset. For the *k-NN* model, the *k* value can be set at the square root of the sample number.

4 Notes

1. One of the largest obstacles to overcome during the analysis of cytological smears is the presence of hemoglobin, a strong ion suppressant. Given the nature of the sampling procedure, it is often difficult to avoid blood being drawn along with the cellular material and can be interspersed throughout the sample.

There are a number of options that can be employed to minimize its effect on the MALDI-MSI analysis:

- Additional washes with high concentrations of Ethanol can be performed in order to remove any excess Hemoglobin that may be sitting on the surface of the sample
- Commonly, the regions of the smear where high amounts of hemoglobin are present can be easily identified in the scanned image due to the dark red/orange appearance. Thus, these regions of the smear can either be avoided during the analysis or virtual microdissection can be performed post-analysis. The effect of virtual microdissection on cytological smears containing hemoglobin can be seen in Fig. 2, with a more complex protein profile being produced along with greater signal intensity of the less abundant proteins.

2. Another troublesome aspect of such cytological specimens is the presence of particularly dense colloidal material in the smeared sample. This type of material is difficult to penetrate with the laser and, as such, little proteomic information can be obtained from these regions. Furthermore, colloidal material can detach during the analysis and ultimately lead to contamination in the mass spectrometer. Therefore, the specimen

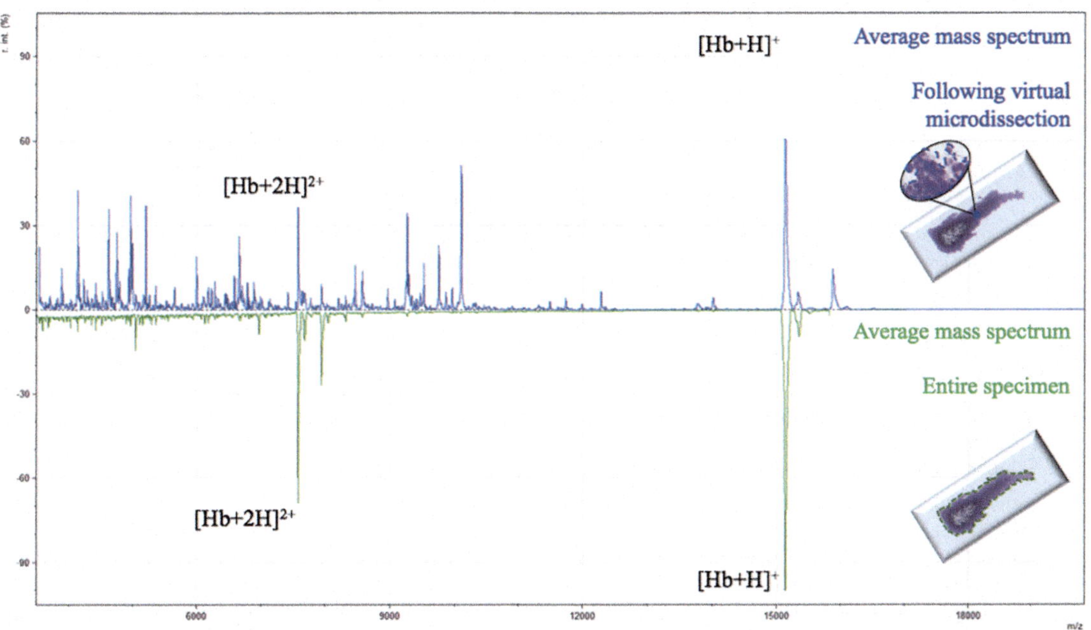

Fig. 2 Average mass spectrum of entire cytological specimen (*Bottom; Green*) and virtually microdissected ROI following annotation from the pathologist (*Top; Blue*). The presence of singly and doubly charged hemoglobin is indicated

should be carefully examined prior to performing the tissue wash, and observed at an oblique angle, in order to identify any colloidal regions of the smear. In cases where large colloid aggregates are present, more aggressive agitation should be performed during the tissue washes in order to remove the excess material prior to the sample entering the MALDI instrument.

3. The dimensions of cytological thyroid smears can often be upward of, if not greater than, 3 cm × 2 cm. Therefore, high spatial resolution images of these specimens can require lengthy image acquisitions, and such analysis times (upward of 12 h) can lead to high levels of instrument contamination and a gradual drop in sensitivity during the acquisition. Ultimately, a compromise between the selected rastering and analysis time needs to be found.

   MALDI-MS Images obtained with 80 μm are sufficient to visualize the cellular aggregates that are present within the smear. However, in some cases, it is not feasible to analyze the entire smear at this rastering. Therefore, large smears may be analyzed with a rastering of between 100 and 150 μm in order to control the acquisition time. However, the capability to obtain specific protein profiles from only the cellular aggregates (i.e., the specificity of the individual profiles) may be lessened somewhat. Alternatively, a brief profiling of the specimen can be performed prior to the measurement region being assigned. Taking this approach, the regions of the smear where pathologically significant information cannot be extracted (i.e., Colloidal regions with low quality matrix homogeneity or regions of the smear containing high amounts of hemoglobin) can be identified and excluded from the measurement region. Ultimately, the regions of the smear that contain the desired proteomic information can then be analyzed at the desired spatial resolution within a reasonable timeframe.

References

1. Jemal A, Bray F, Center MM, Ferlay J, Ward E, Forman D (2011) Global cancer statistics. CA Cancer J Clin 61(2):69–90
2. Cancer Statistics Review, 1975–2011—SEER Statistics [Internet]. Available from: http://seer.cancer.gov/csr/1975_2011/results_merged/sect_26_thyroid.pdf
3. Patel KN, Shaha AR (2014) Poorly differentiated thyroid cancer. Curr Opin Otolaryngol Head Neck Surg 22(2):121–126
4. Tzen C-Y, Huang Y-W, Fu Y-S (2003) Is atypical follicular adenoma of the thyroid a preinvasive malignancy? Hum Pathol 34(7):666–669
5. Vasko VV, Gaudart J, Allasia C et al (2004) Thyroid follicular adenomas may display features of follicular carcinoma and follicular variant of papillary carcinoma. Eur J Endocrinol Eur Fed Endocr Soc 151(6):779–786
6. American Thyroid Association (ATA) Guidelines Taskforce on Thyroid Nodules and Differentiated Thyroid Cancer, Cooper DS, Doherty GM et al (2009) Revised American Thyroid Association management guidelines for patients with thyroid nodules and differentiated thyroid cancer. Thyroid Off J Am Thyroid Assoc 19(11):1167–1214

7. La Vecchia C, Malvezzi M, Bosetti C et al (2015) Thyroid cancer mortality and incidence: a global overview. Int J Cancer J Int Cancer 136(9):2187–2195
8. Polyzos SA, Anastasilakis AD (2009) Clinical complications following thyroid fine-needle biopsy: a systematic review. Clin Endocrinol (Oxf) 71(2):157–165
9. Mainini V, Pagni F, Garancini M et al (2013) An alternative approach in endocrine pathology research: MALDI-IMS in papillary thyroid carcinoma. Endocr Pathol 24(4): 250–253
10. Pagni F, Mainini V, Garancini M, et al (2015) Proteomics for the diagnosis of thyroid lesions: preliminary report. Cytopathol Off J Br Soc Clin Cytol 26(5):318–324
11. Pagni F, De Sio G, Garacini M et al (2016) Proteomics in thyroid cytopathology: relevance of MALDI-imaging in distinguishing malignant from benign lesions. Proteomics 16(11–12):1775–1784

Chapter 6

Droplet-Based Liquid Extraction for Spatially-Resolved Microproteomics Analysis of Tissue Sections

Maxence Wisztorski, Jusal Quanico, Julien Franck, Benoit Fatou, Michel Salzet, and Isabelle Fournier

Abstract

Obtaining information on protein content while keeping their localization on tissue or organ is of importance in different domains to understand pathophysiological processes. There is increasing interest in studying the microenvironment and heterogeneity of tumors, which currently is difficult with existing proteomics techniques. The advent of new techniques, like MALDI Mass Spectrometry Imaging, made a significant progress in the last decade but is characterized by a number of inherent drawbacks. One of these is the limited identification of proteins. New alternative approaches such as spatially-resolved liquid microextraction have recently been proposed to overcome this limitation. In this chapter, we describe strategies using liquid microjunction to perform extraction of previously digested peptides or of intact proteins from tissue section in a localized manner.

Key words Liquid microjunction, LESA, Microproteomics, Protein microextraction, Surface sampling, MALDI imaging

1 Introduction

Through numerous developments including improvement of sample preparation [1–4], unlocking of the commonly conserved clinical tissue samples [5, 6], and computational methods for data analysis [7–9], MALDI Mass Spectrometry Imaging (MSI) has become a powerful tool to obtain information on the precise molecular distribution of endogenous or exogenous compounds on a tissue section. Developments in MALDI-MSI now permit excellent lateral and spectral resolution. For small compounds, identification of molecules is possible directly on the tissue section. In the case of proteins, this question remains a challenge. Several strategies have been proposed like bottom-up [6, 10] or top-down [11, 12] strategies applied directly on the tissue section. These strategies lead to the identification of a small number of proteins that are predominantly the most abundant ones [13]. This could

Laura M. Cole (ed.), *Imaging Mass Spectrometry: Methods and Protocols*, Methods in Molecular Biology, vol. 1618,
DOI 10.1007/978-1-4939-7051-3_6,

be sufficient in finding new biomarkers but not for a deeper understanding of pathophysiological processes. There is increasing interest in studying the microenvironment and heterogeneity of tumors for understanding the complexity of a pathology. Making a precise picture of the disease needs an efficient microsampling method coupled an accurate localization for performing region-dependent proteomics. In another hand, mass spectrometry-based proteomics allows the identification of thousands of proteins and the possibility to quantify them [14, 15]. In this case, most of the time, extraction is performed on organs or small pieces of tissues and these gross tissue extraction treatments lead to loss of the localization of molecules and also possible dilution effects due to the presence of abundant proteins. In this context, we proposed, several years ago, strategies combining MALDI-MSI as support to obtain molecular regions of interest (ROIs) and localized mass spectrometry-based proteomics [16–22]. To perform spatially resolved microproteomics analysis, we developed two strategies, the Parafilm-assisted microdissection and a droplet-based liquid extraction. The first one consists of realizing manual microdissection of ROIs on tissue sections mounted on a parafilm M-covered glass slide, followed by protein extraction, digestion, and analysis of the peptides by nanoLC-MS/MS [19, 21]. The second strategy uses Liquid-Microjunction (LMJ) to perform molecule extraction and could be achieved by two different ways for microproteomics (Fig. 1). First, it is possible to perform in situ microdigestion of ROIs directly on tissue followed by extraction of digested peptides by LMJ [17, 18]. We recently increased the potential of LMJ to perform intact protein extraction from ROIs [22] and used the extract for in-gel or off-gel analysis. In this chapter, we will focus on all the possibilities of the LMJ extraction to realize spatially resolved microproteomics analysis with detailed methods for both microdigestion combined to liquid extraction and intact protein extraction by LMJ.

2 Materials

2.1 Washing Steps for Fresh Frozen Tissue Preparation

1. Ethanol 75% (−20 °C): 75 mL of absolute ethanol (≥99.8%) and water (HPLC grade) to 100 mL. Prepare fresh. Store at −20 °C.
2. Ethanol 95% (−20 °C): 95 mL of absolute ethanol (≥99.8%) and water (HPLC grade) to 100 mL. Prepare fresh. Store at −20 °C.
3. Chloroform (−20 °C): 100 mL of chloroform (≥99.9%). Store at −20 °C. Chloroform is toxic when inhaled, so work in the hood.

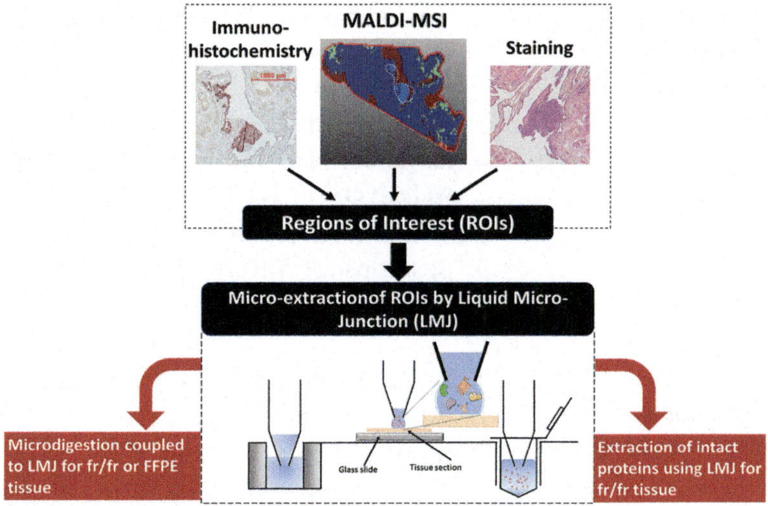

Fig. 1 Schematic representation of strategies used for microproteomics analysis using Liquid MicroJunction (LMJ). The upper panel shows methods to define regions of interest (ROIs) according to highlighted regions, such as by histological staining, immunohistochemistry, or molecular histology profiling by Mass spectrometry Imaging. Once the ROI is defined, microproteomics can be performed in two different ways. Localized on-tissue microdigestion of fresh/frozen (fr/fr) or Formalin-fixed and paraffin-embedded (FFPE) tissue is combined with subsequent liquid microjunction to extract digested peptides. Intact proteins can also be extracted directly from ROIs of fr/fr tissue

2.2 Dewaxing Steps for FFPE Tissue Preparation

1. Xylene: 100 mL of xylene (≥99.9%). Xylene is toxic when inhaled, so work in the hood.
2. Ethanol 100%. Prepare fresh.
3. Ethanol 95%: 95 mL of absolute ethanol (≥99.8%) and water to 100 mL. Prepare fresh.
4. Ethanol 75%: 75 mL of absolute ethanol (≥99.8%) and water to 100 mL. Prepare fresh.
5. Ethanol 30%: 30 mL of absolute ethanol (≥99.8%) and water to 100 mL. Prepare fresh.
6. Water: 100 mL of water (HPLC grade). Prepare fresh.

2.3 On-Tissue Digestion Using a Microspotter

1. Trypsin, sequencing grade modified. Suspend in 50 mM NH_4HCO_3 buffer at 20 μg/mL (*see* **Note 1**).
2. Methanol 50%: 50 mL of absolute methanol and water to 100 mL. Prepare fresh. Methanol is toxic, so work in the hood.
3. Aqueous TFA solution 0.1%. TFA is toxic, so work in the hood.
4. Chemical Inkjet Printer (*see* **Note 2**).

2.4 Modification on Advion Triversa Nanomate with LESA Option

1. Advion Triversa Nanomate platform (Advion BioSciences Inc., Ithaca, NY, USA), with Liquid Extraction Surface Analysis (LESA) option.
2. Chipsoft software (version 8.3.1.1018 build 101018) with the Advance User Interface (AUI) enabled and LESA Points software (version 1.1.0.0).
3. A modified LESA Universal Adapter Plate with predefined positions for solvent and sampling tubes (*see* **Note 3**) [22].
4. Non-conductive tips, 0–10 μL.

2.5 Liquid Microextraction of Digested Peptides

1. Aqueous TFA solution: 0.1% TFA is toxic, so work in the hood.
2. Methanol 70%: 7 mL of absolute methanol and water to 10 mL. Prepare fresh. Methanol is toxic, so work in the hood.
3. Acetonitrile 80%: 8 mL of Acetonitrile and water (HPLC grade) to 10 mL. Prepare fresh. Acetonitrile is toxic, so work in the hood.

2.6 Liquid Microextraction of Intact Proteins

1. SDS Solution: Resuspend SDS in Tris–HCl (pH 8, 0.1 M) with DTT (50 mM) to obtain a final concentration of 1% (w/v).
2. CHAPS Solution: Resuspend CHAPS in Tris–HCl (pH 10, 0.1 M) with DTT (50 mM) to obtain a final concentration of 4% (w/v).

2.7 Shotgun Analysis

1. Urea solution: Resuspend urea in Tris–HCl (pH 8.5, 0.1 M) to obtain a final concentration of 8 M.
2. Microcon filter units (Millipore) YM-30 (*see* **Note 4**).

2.8 Protein Identification and Quantification

1. MaxQuant (latest version is available at: http://www.coxdocs.org/).
2. UniProt reference protein database for the organism of interest.

2.9 Quantification-Based Mass Spectrometry Profiling

1. Perseus Framework (latest version is available at: http://www.coxdocs.org/).
2. TIGR Multiexperiment viewer (MEV v4.9).

3 Methods

This methodological part is decomposed according to the workflow presented in Fig. 1. We present here only methods for microdigestion and/or LMJ. MALDI MSI is not the principal aspect of this method description. Detailed protocols for tissue preparation for MALDI MSI could be found in our previously published chapter for Formalin Fixed, Paraffin embedded (FFPE), or Fresh and frozen (fr/fr) tissue sections [23]. Complementary to data obtained from MALDI-MSI, information from histology or immunohistology could be used to determine regions of interest. Concerning MALDI-MSI, different algorithms are used to realize spatial segmentation of spectra in order to obtain molecular histology [24]. In this case, regions are defined only based on their molecular signatures and not on histological information like the morphology of cells for example. Molecularly different ROIs are selected and localized microproteomics is performed on these regions.

We primarily describe the steps leading to the preparation of tissue sections prior to microproteomics analysis, which is an extremely important part of the method. For fresh frozen tissue sections, an important first step is the washing of the sample, like in MALDI MSI. The use of an ethanol bath for fixation of proteins and removal of endogenous salts and peptides is coupled to a wash using chloroform for depletion of lipids. These washes prevent the extraction of these components as they could interfere with subsequent analysis after microdigestion. In the case of the extraction of intact proteins, washing the fresh frozen tissue section is also crucial. Indeed, LMJ using detergents, which act like a "liquid microdissection," removed the totality of the tissue at the junction point. Complete removal of the tissue is observed in the case of the washed section, contrary to the unwashed one. Without washing steps, a severe decrease of the number of proteins identified is observed [22]. In the case of trypsin digestion (and for intact protein extraction in fact), the different washing steps destabilize the lipid bilayer membrane of the cells allowing a better penetration efficiency for the trypsin or for the detergent.

For FFPE tissue sections, in addition to the classical paraffin removal steps, an antigen retrieval step has to be performed. This step allows for better enzyme accessibility of the cleavage sites and significantly increases the efficiency of digestion. Antigen retrieval is very specific to the tissue composition and needs to be optimized and chosen wisely. Fixation using paraformaldehyde prevents extraction of intact proteins due to crosslinks formed between molecules. The only way in this case is to use a microdigestion step prior to extraction.

Concerning the parameters for microdigestion, the use of an automatic microdispenser that is accurate in terms of locating the dispense position and precise in terms of repeatability of deposition is important. A chemical inkjet printer permits this type of precision. In this case, the digested region will delimit the analysis region so that even if larger microextracted areas are achieved with LMJ, the area where it will extract digested peptides will remain defined by the microdigested region. This allows the analysis of features smaller than 300 μm using a chemical inkjet printer for trypsin

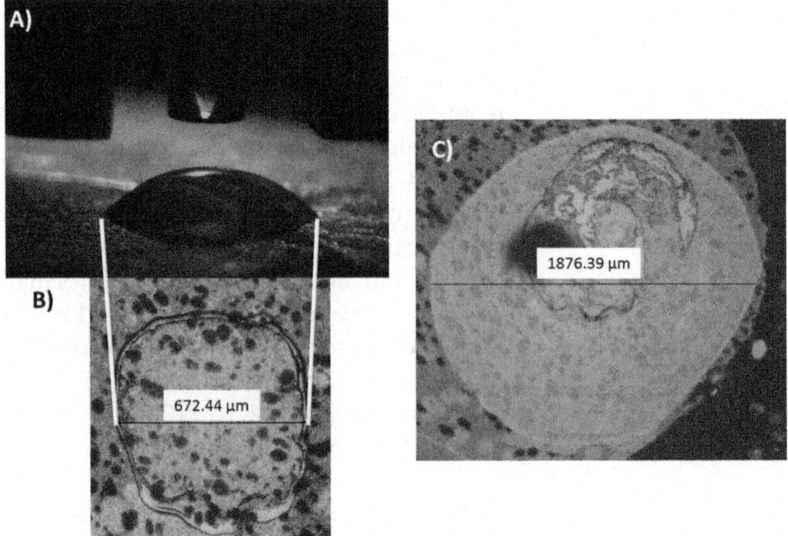

Fig. 2 (**a**) Optical image of a microdigestion droplet maintained on the surface of a tissue section. (**b**) Magnified view of the digested region and (**c**) the same region after LMJ. The previously digested region is observed inside the extracted region of 1.8 mm in diameter

deposition. The volume and number of the droplets are also important points to optimize. In the case of microdigestion for a microproteomics experiment, the digestion needs to be controlled by maintaining the droplet formed on the surface of the tissue. Figure 2a shows the droplet of solution containing trypsin formed on the surface of the tissue section. For this example, the digested region is 672 μm in diameter (Fig. 2b). If the volume of the droplet is not enough, the efficiency of digestion will be decreased but if the droplet size becomes too large, an increase of the diameter of the region will lead to a less precise analysis. These parameters are dependent on the composition of the tissue and need to be adapted for each situation. If the droplet is correctly maintained during 2 h, no additional incubation time is needed. As observed in Fig. 2b, the digested region is well defined and no spread is observed. Figure 2c shows the previously digested region after LMJ. The size of the extracted region is more than 1.5 mm in diameter (1876 μm). In this case, the size of the analysed region is defined by the size of the digested region, as shown previously [17].

Concerning liquid extraction, the size of the microjunction is defined by several parameters, such as the volume of solution used to create the junction and the distance between the surface and the end of the collecting tip. These parameters need to be tested properly before the extraction process. Typically, a distance of dispense between 0.2 and 0.4 mm and a volume between 0.6 and 0.9 μL will lead to a junction of around 1 mm in diameter. Figure 3

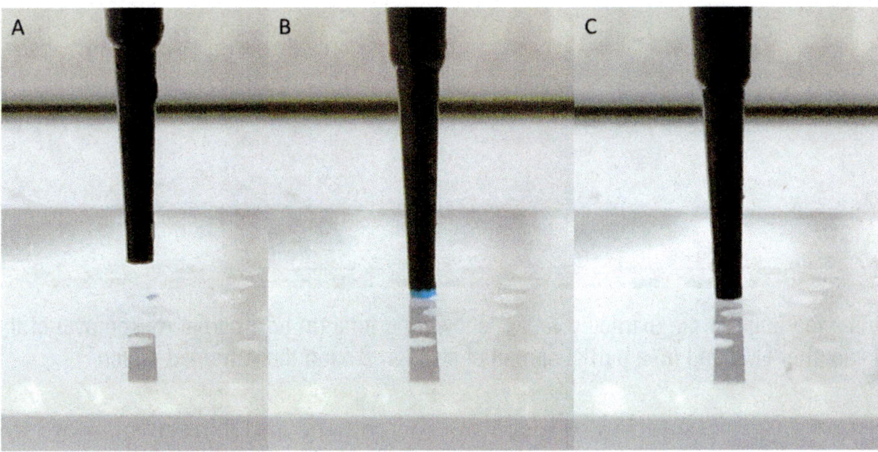

Fig. 3 Photograph showing varying distances between the conical tip and the sample. (**a**) The dispensing end of the tip is too far from the surface and no junction is formed. (**b**) Optimal distance between the tip end and the surface with formation of a liquid microjunction (blue liquid used for a better visualization). (**c**) The tip end is too close to the surface and the contact between the liquid and the tissue is limited

presents the three possibilities for the setting of the distance between the tip and the surface to analyze. In Fig. 3a, the tip is too high and no liquid is in contact with the surface. On the other hand, in Fig. 3c, the tip is too close and the junction is not visible. In this case, extraction could occur but the amount of solution is not enough to be able to do optimal extraction. In Fig. 3b, the distance and the volume are optimal to maintain a microjunction. A repeat of the extraction cycle could be done. For the extraction of digested peptides, it has been observed that successive cycles using a combination of different solutions will increase the extraction efficiency. For intact proteins, repeating the number of cycles realized in the same position is also possible. However, too many repeats will increase the size of the junction and decrease the precision of the analysis. Three cycles of extraction are normally enough to remove all the tissue in the extracted region and if some tissue remained, an additional extraction step can be performed. Figure 4 shows three scenarios that could be observed after liquid microjunction. In Fig. 4a, total removal of the tissue is observed and this corresponds to the optimal settings. Figure 4b presents a partial removal of the tissue. Optimization is needed either by choosing a more appropriate detergent or by increasing the number of extraction cycles. If spreading of liquid is observed (Fig. 4c), the dispensed volume needs to be adjusted or the number of extraction cycles is too much.

Concerning the solution used to perform extraction after microdigestion, various solvents are employed. Combining three consecutive extractions with different solvents significantly improves the extraction. Optimizations have been done using first an acidic

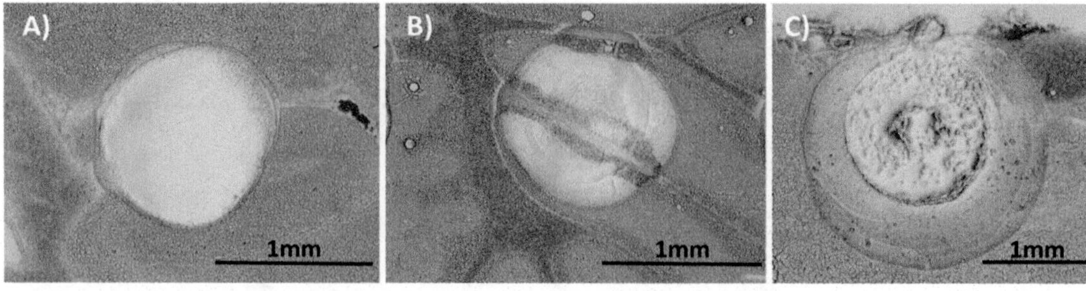

Fig. 4 Photomicrographs of the extracted region showing a total (**a**) or a partial (**b**) removal of the tissue or spread of liquid after LMJ and (**c**) a partial spread of solution around the extracted region

solution to inactivate the enzyme and extract the more hydrophilic peptides. Then, two consecutive extractions using solutions of methanol and acetonitrile were performed to extract all remaining peptides. Extraction of intact proteins is performed using a solution containing detergents like CHAPS or SDS. Solutions containing only organic solvents extract a small fraction of major proteins. Deeper analysis needs the use of a detergent solution to be sure to extract the maximum amount of proteins at the junction.

After LMJ, the extract is dispensed on a collecting tube and could be used for various analyses that can be found in previously published studies [17, 18, 22].

3.1 Preparation of Fresh/Frozen Tissue Section

1. Tissue sections with a thickness of 20 μm are used (*see* **Note 5**) and mounted sections are stored in a sealed container at −80 °C until use.
2. Before use, slides are warmed at room temperature under vacuum to prevent water condensation on the surface of the frozen slide.
3. After complete drying, the glass slide is washed using different solutions. The washing step is a critical point and skipping this step could lead to a decrease of the number of proteins extracted.
4. The tissue section is immersed first in the 75% ethanol solution for 30 s. No agitation or shaking is needed. This step washes out salts, cell fragments, or residual fluids.
5. The glass slide is then dried under vacuum during 5 min.
6. After complete drying, the glass slide is dipped into the 95% ethanol solution for 30 s. No agitation or shaking is needed. This step prevents degradation of the proteome by dehydration and fixation of the tissue.
7. The glass slide is again completely dried during 5 min.
8. After complete drying, immerse the glass slide gently in chloroform during 30 s. No agitation or shake is needed. This step

removes lipids (especially phospholipids) present in high concentrations in the tissue (components of cell membranes).

9. Take the slide out and place it in a vacuum desiccator for complete drying of the tissue.

3.2 Preparation of FFPE Tissue Section

1. The glass slide with the tissue section (20 μm thickness) is dried overnight at 40 °C for good adherence.
2. Dewaxing is performed by gently dipping the glass slide in a xylene both for 5 min. This procedure is repeated twice. No agitation or shaking is needed.
3. The tissue section is then rehydrated by stepwise immersion in 100% ethanol twice, 95% ethanol, 75% ethanol, and 30% ethanol, for a duration of 5 min each time.
4. The glass slide is placed in the vacuum desiccator to dry the sections completely.
5. In most of the cases, antigen retrieval steps are needed to obtain better efficiency during trypsin digestion.

3.3 Extraction of Digested Peptides from Fresh/Frozen or FFPE Tissue

3.3.1 On-Tissue Microdigestion Using a Microspotter

1. A microspotter, like the Chemical Inkjet Printer CHIP-1000 (Shimadzu Biotech, Kyoto, Japan), can be used. CHIP-1000 is a piezoelectric solvent delivery system able to deposit picoliter volumes of reagents on defined locations with a precision of less than 5 μm. The great advantage of using this type of device is the possibility to digest regions with diameters from 250 μm up to 1 mm with high reproducibility and constant quantity of enzyme.
2. The glass slide subjected to prior washing steps for fresh frozen or dewaxing for FFPE tissue is used.
3. To improve the efficiency of digestion, settings of the CHIP-1000 concerning the number of spots per ROI, the distance between each spot in a region, the volume of enzyme deposited at each spot per cycle, and the time between each cycle are defined to maintain a droplet on the surface of the tissue.
4. For example, for a square region with an area of $1 mm^2$, a pattern of 4 × 4 spots is designed with a distance between spots of 200 μm is used. A total droplet volume of 100 pL/spot/cycle is deposited.
5. The enzyme solution in the microdispenser is replaced with a fresh aliquot every 30 min to minimize autolysis.
6. The cyclic deposition on the predefined square pattern is repeated for a duration of 2 h.
7. After enzyme deposition, a solution of aqueous 0.1% TFA is spotted for 25 cycles to stop the digestion (*see* **Note 6**).

3.3.2 Liquid Microextraction

1. Eppendorf tubes with different solvents are placed on predefined positions in the LESA Universal Adapter Plate (*see* **Note 7**).
2. The solution of TFA is used as the first extraction solution.
3. The robotic arm first picks up a disposable tip, then aspirates, a volume of 1.7 μL of the solution of TFA from the corresponding solvent tube.
4. The robotic arm then moves the tip above the ROI to be extracted, and descends on the preset height.
5. The distance between the slide and the tip is specified between 0.2 mm and 0.4 mm above the tissue section (*see* **Note 8**).
6. A volume of 0.8 μL is dispensed to form the liquid microjunction (*see* **Note 9**).
7. The liquid microjunction is maintained during 20 s.
8. A volume of 0.9 μL is aspirated and the tip is moved to the collecting tube.
9. **Steps 2–7** are repeated with solutions of methanol and acetonitrile consecutively (*see* **Note 10**).
10. Each time, the extraction solution is mixed in the same collecting tube.
11. After extraction, the collecting tube is dried under vacuum.
12. The digested peptides are reconstituted in 20 μL of 5% ACN/TFA 0.1% and are directly used for shotgun proteomics analysis.

3.4 Intact Protein Analysis

3.4.1 Extraction of Intact Proteins from Fresh/Frozen Tissue

1. Eppendorf tubes containing different solvents are placed on predefined positions in the LESA plate.
2. A solution of SDS or CHAPS is used.
3. After getting a tip, a volume of 2 μL of the detergent solution is aspirated.
4. The tip is moved above the ROI to be extracted.
5. The distance between the slide and the tip is specified between 0.2 mm and 0.4 mm above the tissue section.
6. A volume of 0.8 μL is dispensed to form the liquid microjunction (*see* **Note 11**).
7. The liquid microjunction is maintained for 20 s.
8. A volume of 1 μL is aspirated and the tip is moved to the collecting tube.
9. **Steps 2–7** are repeated with the same solution twice.
10. The extracted solution is mixed in the same collecting tube.
11. The extract solution could be directly loaded on an SDS-PAGE, for western-blot analysis or the proteins can be recovered using the FASP protocol [22].

3.4.2 Shotgun Analysis

1. A sonication step to shear DNA followed by centrifugation at 14,000 × *g* for 5 min to clarify the solution reduces extract viscosity and avoids clogging the filter.
2. A filter with a cutoff of 30 kDa is used (*see* **Note 4**).
3. CHAPS detergent solution seems to be more suitable for microproteomics conditions employing the FASP protocol because CHAPS forms micelles of smaller sizes compared to SDS (6 kDa for CHAPS and 18 kDa for SDS).
4. The steps involved in the FASP protocol have been described previously [25]. Mass spectrometry-based shotgun proteomics analysis is then realized on the extracts.

3.5 Protein Identification and Label-Free Quantification

1. Separation and tandem MS/MS of peptides are performed using an online reverse-phase chromatographic system interfaced to a Thermo Scientific Q Exactive mass spectrometer.
2. MS/MS spectra are processed by database matching using the Andromeda search engine [26] for peptide and protein identification, and relative label-free quantification of proteins is done using the MaxLFQ algorithm [27] altogether integrated into MaxQuant (*see* **Note 12**).
3. The appropriate database is downloaded from UniProt [28].
4. The FDR at the peptide and protein levels were set to 1%.

3.6 Comparison of Different Regions

1. To statistically compare regions, several extractions on the same region are performed. If the region is large enough, extractions are realized on the same tissue section. If the region is too small, extractions are performed on consecutive tissue sections.
2. Analysis of identified proteins is achieved using Perseus software [29] in order to filter out hits to the reverse database, proteins only identified with modified peptides, and potential contaminants.
3. The LFQ intensities are log2-transformed.
4. Extraction points of the same region are grouped and statistical analyses like t-test or multiple samples tests like ANOVA are performed.
5. The resulting proteins with significantly different LFQ intensities compared to the control are used for hierarchical clustering to visualize clusters of the data after *Z*-score normalization.
6. The over- or under-expressed proteins are obtained and used for further analysis (Overrepresentation and protein/gene enrichment pathway analysis, for example).

3.7 Quantification-Based Mass Spectrometry Profiling

1. To realize quantification-based mass spectrometry profiling, extractions on consecutive points of a tissue section are done.
2. The distance between each point is defined by the size of the analysis regions.
3. For microproteomics analysis using microdigestion and LMJ, the size could be less than 500 μm. In this case, the experiment will have to be done twice with the digested regions spaced 1.5 mm center to center. After microextraction by LMJ, a second series of digestion is performed- with a shift of 750 μm relative to the previous digestion regions. In this case, distance between two points in a series is enough to prevent overlapping extractions due to the larger diameter size of LMJ compared to the microdigested area.
4. For intact protein extraction, the minimal distance between two consecutive points is 1.5 mm using conventional nonconductive tips.
5. Analysis of identified proteins is achieved using Perseus software [29] as in Subheading 3.6.
6. LFQ intensities were log-transformed (log2(x)) and exported to TIGR Multiexperiment viewer (MEV v4.9) [30].
7. Root-mean-square (RMS) normalization is done and the values for all points are visualized using a color scale and superimposed with the image of the tissue section as presented in Fig. 5. In this figure, localization of five different proteins is represented in ten regions along a transversal section of a rat brain. The localization of all these proteins (in terms of LFQ abundance) is in correlation with their biological function.

4 Notes

1. Different concentrations of trypsin or other types of enzyme.
2. Other instruments can be used in place of the CHIP-1000, such as a microarray spotter.
3. A conventional 386-well plate can be used. In this case, the glass slide needs to be secured on the plate with double-sided tape.
4. Due to the concentration of detergent used in this case, the possibility of forming micelles is high. Filter membranes with a cutoff smaller than 30 kDa will not efficiently remove detergent micelles and lead to a decrease of the digestion efficiency.
5. Optimization of the methods was done using sections 20 μm thick to obtain enough proteins for analysis. Thinner tissue sections can be used and extracts from the same region on consecutive tissue sections can be pooled for analysis.

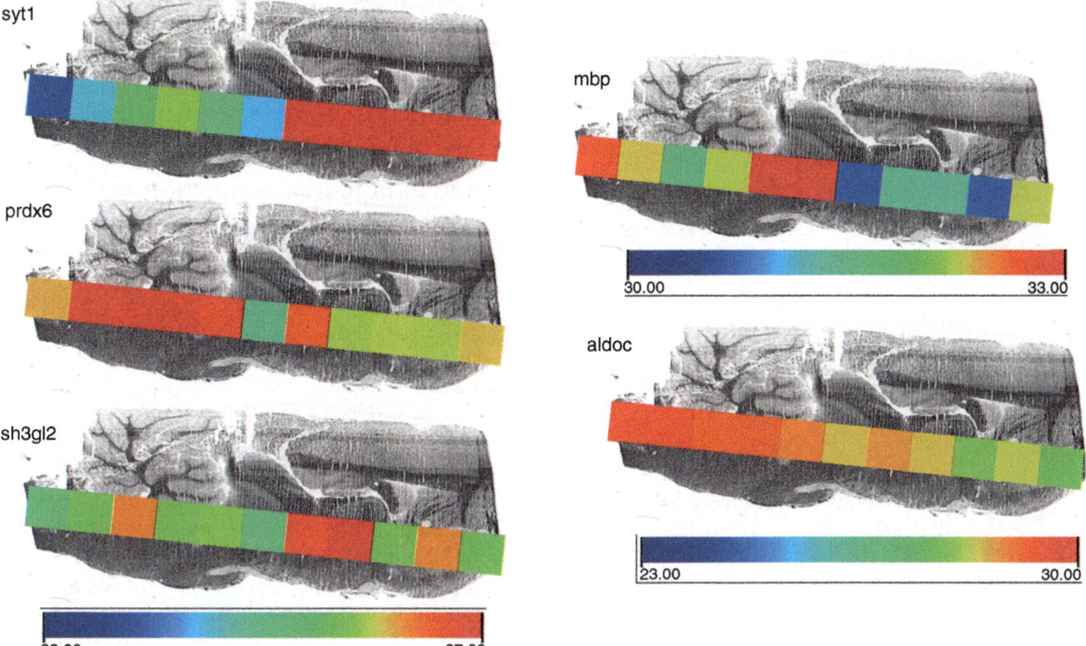

Fig. 5 Localization of five proteins according to their relative quantification values on ten points of a rat brain tissue section. Syt1: Synaptotagmin 1, prdx6: Peroxiredoxin-6, sh3gl2: Endophilin-A1, mbp: Myelin Basic Protein, aldoc: Aldolase C

6. This step is also acting as a conditioning step to facilitate the extraction of tryptic peptides.

7. Tubes containing volatile solvents need to be sealed with a piece of aluminum foil to avoid solvent evaporation. At the beginning of the solvent aspiration step, the tip will pierce through the foil.

8. The distance needs to be optimized for each type of tissue and for the solvent used to obtain the smallest size of liquid microjunction. According to the previously published experiments by Kertesz V, et al. [31], the diameter of extraction obtained with solutions containing a low amount of organic solvent will be around 1 to 2 mm. Using solutions with high concentrations of methanol or acetonitrile will yield liquid microjunctions with diameters between 2 and 3 mm.

9. The goal here is to maintain a liquid microjunction and not to deposit a droplet of solvent. The microjunction will limit the spread of the solvent thus producing smaller analyzed regions compared when the droplet is merely deposited on the tissue section. In the case of the extraction of peptides from a region previously digested, previous experiments have shown that using an appropriate solvent, extraction is limited to tryptic peptides and that the size of the digested region will define the

size of the analyzed region. Using a microspotter, a region of around 500 μm in diameter can be easily analyzed.

10. The volume of solution of methanol and acetonitrile used to realize the liquid microjunction is decreased at 0.6 μL to reduce the size of the microjunction.
11. The goal here is to maintain liquid microjunction and not to deposit a droplet of solvent. The microjunction will limit the spread of the solution compared to the deposition of a droplet.
12. Other proteomics identification and quantification tools can be used.

References

1. Crecelius AC, Schubert US, von Eggeling F (2015) MALDI mass spectrometric imaging meets "omics": recent advances in the fruitful marriage. Analyst 140:5806–5820
2. Goodwin RJA (2012) Sample preparation for mass spectrometry imaging: small mistakes can lead to big consequences. J Proteome 75:4893–4911
3. Chaurand P (2012) Imaging mass spectrometry of thin tissue sections: a decade of collective efforts. J Proteome 75:4883–4892
4. Lemaire R, Wisztorski M, Desmons A, Tabet JC et al (2006) MALDI-MS direct tissue analysis of proteins: improving signal sensitivity using organic treatments. Anal Chem 78:7145–7153
5. Oetjen J, Lachmund D, Palmer A, Alexandrov T et al (2016) An approach to optimize sample preparation for MALDI imaging MS of FFPE sections using fractional factorial design of experiments. Anal Bioanal Chem 408:6729–6740
6. R. Lemaire, A. Desmons, J. C. Tabet, R. Day, et al., Direct analysis and MALDI imaging of formalin-fixed, paraffin-embedded tissue sections J Proteome Res 2007 6, 1295–1305.
7. Cassese A, Ellis SR, Ogrinc Potočnik N, Burgermeister E et al (2016) Spatial autocorrelation in mass spectrometry imaging. Anal Chem 88:5871–5878
8. Alexandrov T, Chernyavsky I, Becker M, von Eggeling F, Nikolenko S (2013) Analysis and interpretation of imaging mass spectrometry data by clustering mass-to-charge images according to their spatial similarity. Anal Chem 85:11189–11195
9. Trim PJ, Djidja M-C, Muharib T, Cole LM et al (2012) Instrumentation and software for mass spectrometry imaging—making the most of what you've got. J Proteome 75:4931–4940
10. Groseclose MR, Andersson M, Hardesty WM, Caprioli RM (2007) Identification of proteins directly from tissue: in situ tryptic digestions coupled with imaging mass spectrometry. J Mass Spectrom 42:254–262
11. Ait-Belkacem R, Berenguer C, Villard C, Ouafik L et al (2014) MALDI imaging and in-source decay for top-down characterization of glioblastoma. Proteomics 14:1290–1301
12. Debois D, Bertrand V, Quinton L, De Pauw-Gillet M-C, De Pauw E (2010) MALDI-in source decay applied to mass spectrometry imaging: a new tool for protein identification. Anal Chem 82:4036–4045
13. Schober Y, Schramm T, Spengler B, Römpp A (2011) Protein identification by accurate mass matrix-assisted laser desorption/ionization imaging of tryptic peptides. Rapid Commun Mass Spectrom 25:2475–2483
14. Beck M, Schmidt A, Malmstroem J, Claassen M et al (2011) The quantitative proteome of a human cell line. Mol Syst Biol 7:549
15. Nagaraj N, Wisniewski JR, Geiger T, Cox J et al (2011) Deep proteome and transcriptome mapping of a human cancer cell line. Mol Syst Biol 7:548
16. Lemaire R, Desmons A, Tabet JC, Day R et al (2007) Direct analysis and MALDI imaging of formalin-fixed, paraffin-embedded tissue sections. J Proteome Res 6:1295–1305
17. Quanico J, Franck J, Dauly C, Strupat K et al (2013) Development of liquid microjunction extraction strategy for improving protein identification from tissue sections. J Proteome 79:200–218
18. Wisztorski M, Fatou B, Franck J, Desmons A et al (2013) Microproteomics by liquid extraction surface analysis: application to FFPE tissue to study the fimbria region of tubo-ovarian cancer. Proteomics Clin Appl 7:234–240
19. Franck J, Quanico J, Wisztorski M, Day R et al (2013) Quantification-based mass spectrome-

try imaging of proteins by parafilm assisted microdissection. Anal Chem 85:8127–8134

20. Fatou B, Wisztorski M, Focsa C, Salzet M et al (2015) Substrate-mediated laser ablation under ambient conditions for spatially-resolved tissue proteomics. Sci Rep 5:18135
21. Quanico J, Franck J, Gimeno JP, Sabbagh R et al (2015) Parafilm-assisted microdissection: a sampling method for mass spectrometry-based identification of differentially expressed prostate cancer protein biomarkers. Chem Commun (Camb) 51:4564–4567
22. Wisztorski M, Desmons A, Quanico J, Fatou B et al (2016) Spatially-resolved protein surface microsampling from tissue sections using liquid extraction surface analysis. Proteomics 16:1622–1632
23. Wisztorski M, Franck J, Salzet M, Fournier I (2010) MALDI direct analysis and imaging of frozen versus FFPE tissues: what strategy for which sample? Methods Mol Biol 656:303–322
24. Walch A, Rauser S, Deininger S-O, Höfler H (2008) MALDI imaging mass spectrometry for direct tissue analysis: a new frontier for molecular histology. Histochem Cell Biol 130:421–434
25. Wiśniewski JR, Zougman A, Nagaraj N, Mann M (2009) Universal sample preparation method for proteome analysis. Nat Methods 6:359–362
26. Cox J, Neuhauser N, Michalski A, Scheltema RA et al (2011) Andromeda: a peptide search engine integrated into the MaxQuant environment. J Proteome Res 10:1794–1805
27. Cox J, Hein MY, Luber CA, Paron I et al (2014) Accurate proteome-wide label-free quantification by delayed normalization and maximal peptide ratio extraction, termed MaxLFQ. Mol Cell Proteomics 13:2513–2526
28. The UniProt Consortium (2012) Reorganizing the protein space at the Universal Protein Resource (UniProt). Nucleic Acids Res 40: D71–D75
29. Tyanova, S., Temu, T., Sinitcyn, P., Carlson, A., et al., 2016 The Perseus computational platform for comprehensive analysis of (prote) omics data. Nat Methods, 13:731–740
30. Howe E, Holton K, Nair S, Schlauch D et al (2010) Mev: multiexperiment viewer. Biomed Inform Cancer Res:267–277
31. Kertesz V, Van Berkel GJ (2014) Sampling reliability, spatial resolution, spatial precision, and extraction efficiency in droplet-based liquid microjunction surface sampling. Rapid Commun Mass Spectrom 28:1553–1560

Chapter 7

DESI Mass Spectrometry Imaging (MSI)

Emmanuelle Claude, Emrys A. Jones, and Steven D. Pringle

Abstract

Desorption Electrospray Ionization (DESI) mass spectrometry is a technique that allows chemical information to be obtained directly from a wide range of surfaces. Using a 2D stage, DESI can be implemented in an imaging mode whereby MS spectra are collected by rastering the spray across the whole surface. Here, we describe the implementation and optimization of DESI imaging for metabolites and lipids from tissue sections using oa-TOF mass spectrometers.

Key words DESI, MS, MSI, Imaging, Lipids

1 Introduction

Desorption Electrospray ionization (DESI), first introduced by Cooks et al. [1] in 2004, has emerged as a powerful and versatile ambient ionization surface analysis technique. Molecules are desorbed and ionized directly from the surface under ambient conditions and analyzed by mass spectrometry. The removal of the sample preparation constraint and requirement for the sample to be introduced to the vacuum regions of the mass spectrometer differentiates DESI from other mainstream MS imaging modalities, opening up a range of novel applications while greatly simplifying the analysis.

A wide range of molecules including polymers [2], pharmaceutical compounds [3], lipids [4], glycans [5], peptides [6], and proteins [7] have been shown to be detectable directly from different sample surfaces. Particular attention has focused on biological tissues; where it has been demonstrated in the last few years that the DESI technique can be used for the detection of a multitude of endogenous lipids that are correlated with biochemical processed and also with disease and disease progression [4, 8, 9]. With a host of recent publications indicating that the phospholipids signature can be used to classify tissue types, DESI-MS shows significant

Laura M. Cole (ed.), *Imaging Mass Spectrometry: Methods and Protocols*, Methods in Molecular Biology, vol. 1618,
DOI 10.1007/978-1-4939-7051-3_7,

potential to be a powerful diagnostic technique in the general area of oncology [10].

The DESI technique consists of charged solvent droplets carried by a nebulizing gas, impacting upon a surface, thereby desorbing and ionizing molecules present on the surface that are then transported toward the atmospheric pressure inlet of the mass spectrometer. Rastering the DESI sampling spray over a sample and assigning coordinates to each mass spectrum collected at a particular position allows chemical maps of the surface to be created and retrospectively interrogated; thus no prior knowledge of the sample is required and the technique can be used as an untargeted imaging approach.

Here, we describe the workflow to obtain DESI-MS image maps of small molecules metabolites and lipid signals directly from tissue sections (porcine liver and mouse brain) using a Prosolia 2D DESI source, mounted on a hybrid quadrupole orthogonal acceleration time-of-flight (Q-TOF) mass spectrometer (either the Waters Xevo G2-XS or SYNAPT G2-Si).

A critical component of the DESI imaging system is the sprayer. There have been studies reporting poor repeatability and reproducibility of DESI using human oesophageal tissue [11] sections and a large-scale inter-laboratory study using rhodamine-coated slides [12]. Here, we detail the procedure that can be used to optimize the DESI imaging system for the high sensitivity analysis of lipids in both positive and negative ion modes using mouse brain and porcine liver samples with a sprayer specifically designed, by Waters, (Wilmslow, UK), for tissue imaging studies.

2 Materials

2.1 Tissue Preparation

Frozen tissue sections are generated using the same protocol [13] as for other MSI techniques. Typically, tissue is sectioned at a thickness of 10–15 μm using a cryostat-microtome (e.g., CM3050 S, Leica, Buffalo Grove, IL, US). Once cut, sections should be kept in a −80 °C freezer until required (Subheading 3.5 of Chap. 3).

2.2 Reagents

Ideally, fresh solutions should be prepared to provide optimal results.

1. Solvents: Water (ultrapure), Methanol HPLC grade or higher.
2. Red Sharpie containing Rhodamine 6G dye.

2.3 Equipment

1. DESI source is a Prosolia 2D source.
2. Pump 11 Elite infusion only single syringe pump (Harvard Apparatus, Holliston, MA, US).

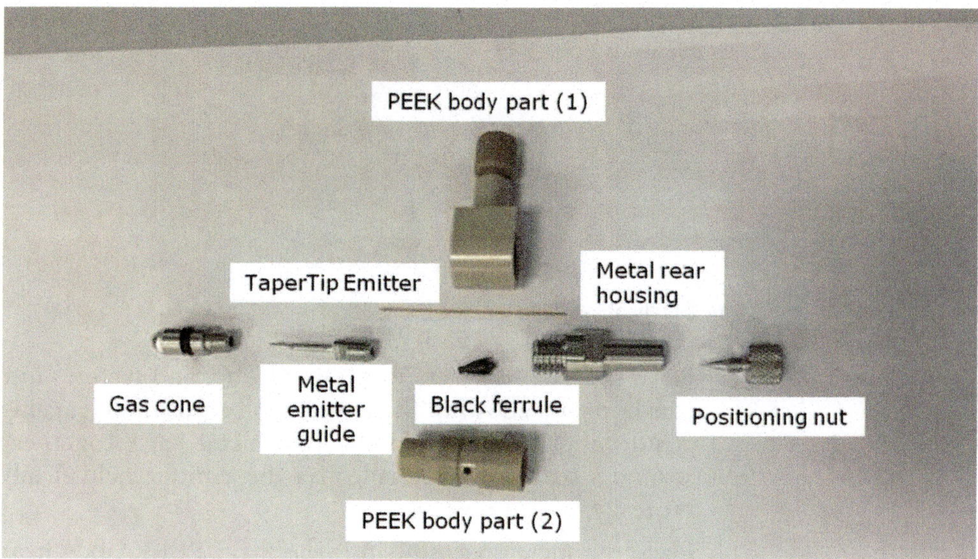

Fig. 1 Parts of the Waters DESI sprayer

3. Mass spectrometer—Xevo G2-XS QTof or the SYNAPT G2-Si HDMS (Waters, Corporation, Milford, MA, US). DESI platforms for other mass spectrometers are also available, or in some instances can be home-built.
4. The DESI sprayer (Waters, Corporation, Wilmslow, UK) contains a 20 μm inner diameter TaperTip capillary (New Objective, Woburn, MA, US) cut to 40 mm length (*see* **Note 1**), a main body consisting of two PEEK parts, a metal gas cone with an aperture of 400 μm diameter, a metal emitter guide and rear housing with a black ferrule, and a positioning metal nut (*see* Fig. 1).

2.4 Software

1. Prosolia Omni Spray 2D: software to control the DESI source for DESI sprayer optimization (Prosolia Inc., Indianapolis, IN, US).
2. MassLynx v4.1: software to control the DESI source for DESI imaging experiments and to control the mass spectrometer (Waters, Corporation, Wilmslow, UK).
3. High Definition Imaging (HDI) 1.4: software to set up DESI imaging experiment, processing, and visualization of the DESI imaging data (Waters, Corporation, Wilmslow, UK).

3 Methods

3.1 Sprayer Construction

1. Screw the positioning nut at the back of metal rear housing.
2. Feed the cut end of the emitter through the metal emitter guide tube and the ferrule.

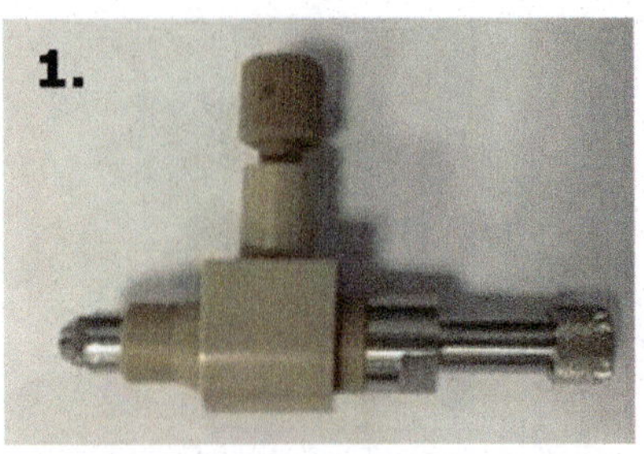

Fig. 2 Fully assembled Water DESI sprayer

3. Place the assembly (emitter, guide tube, and ferrule) into the rear housing, allowing the emitter to be positioned against the positioning nut. Tighten the two metal parts together with spanners so that the ferrule has the emitter held firmly (*see* **Note 2**).
4. Place the metal assembly into the main PEEK body from the back.
5. Carefully place the gas cone to the front of the main PEEK body, avoiding damaging the emitter tip (*see* **Note 3**). Wind the spray cone using a 5 mm spanner until the emitter starts to protrude. The fully assembled sprayer is shown in Fig. 2. Remove the positioning tip from the back of the sprayer.
6. Mount the sprayer assembly using the mounting bracket onto the arm of the DESI source (*see* **Note 4**).
7. Connect the High voltage cable, gas, and solvent fittings (*see* **Note 5**).

3.2 DESI Signal Optimization

A critical aspect to a successful DESI experiment is the production of a focused, co-axially aligned spray. This is governed by the relationship between the solvent flow, nebulizing gas flow, the nozzle protrusion, the nozzle to emitter tip distance, and applied high voltage (*see* Fig. 3). Once the sprayer parameters are optimum the final task is to optimize the geometric parameters, i.e., the sprayer height above the surface and the distance between the sprayer and the inlet capillary.

1. Typical starting positions are described in Table 1.
2. To determine if the system is operational and to obtain suitable starting point for the optimization using tissue samples, draw a large rectangle on a glass slide with a red Sharpie® that contains Rhodamine 6G dye (m/z 443) for positive mode analysis or black permanent marker (m/z 666) for negative mode. Following this preliminary procedure using the ink (*see* **Note 6**), the system needs to be further optimized using a tissue section, ideally one of the same origin as those intended for the study. Take the glass slide with the tissue section mounted from the

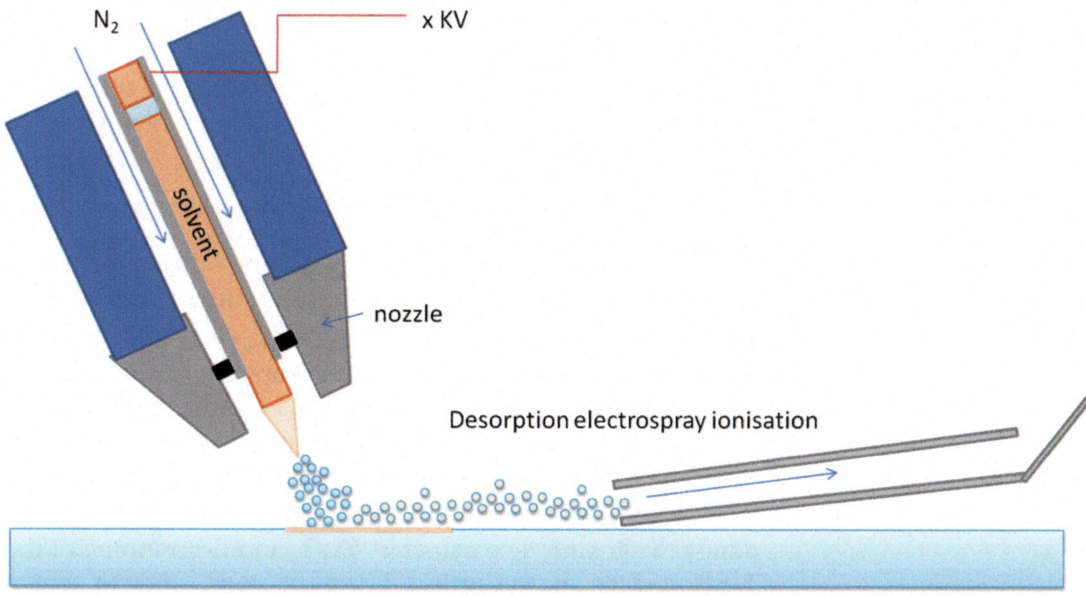

Fig. 3 Schematic of DESI with the Waters sprayer showing the gas, solvent, and high voltage applied onto the sprayer with the capture of the desorbed charged droplets into the metal capillary and into the mass spectrometer API source

Table 1
Starting DESI imaging settings

Height of emitter above surface	1–2 mm
Emitter to inlet distance	5 mm
Sprayer angle	75°
Solvent flow	1 2-μL/min
Gas pressure	5–7 bar
Emitter protrusion	0.2–2 mm
Solvent composition	95:5 methanol water
Voltage	2–5 kV
MS scan time	1 s

−80 °C degrees freezer. Let it defrost at room temperature. No other sample preparation is required. Place the sample onto the DESI stage (*see* **Note 7**).

3. Solvent composition for DESI imaging depends on the analyte of interest and the tissue surface. For lipids in both negative and positive modes 95–98% aqueous methanol is recommended as a starting point (*see* **Note 8**). Many different solvent compositions and additives have been reported [14, 15] and may also be used.

4. The flow of the solvent can be provided either by using the mass spectrometers' internal fluidics or alternatively an external syringe pump or LC pump. It is critical that the system can provide a stable flow rate at the microliter per minute range (*see* **Note 9**).
5. Check an empty region of the glass side to ensure that mass spectral peaks from the solvent are observed. If not, check that solvent is flowing correctly and there are no leakages or blockages in the solvent line or for a damaged emitter (*see* **Note 10**).
6. Place the sprayer above the tissue section (or drawn ink region). Start with the emitter positioned completely inside the gas cone. With a 5 mm spanner, rotate the gas cone in small increment to allow the emitter to protrude out of the gas cone until lipid/ink signals are observed (*see* **Note 11**).
7. After finding the protrusion generating the highest signal, adjust the geometric parameters (X, Y, and Z positions) of the DESI sprayer, the gas pressure (0.5 bar steps), and the source voltage (0.5 kV steps). *See* Fig. 4.
8. A few iterations of **steps 6** and **7** might be necessary.

3.3 DESI Imaging Setup

1. Acquire an optical image (photograph or flat-bed scanner) of the tissue section(s) mounted on a conventional glass slide (*see* **Note 12**). Place the sample onto the 2D–DESI stage.
2. Coregister the optical image using fiducial markers using the High Definition Imaging (HDI) software.
3. Define the rectangular region to be imaged from the coregistered optical image containing the tissue section.
4. Define the pixel size (μm) and the speed of the stage (μm/s). The acquisition rate is calculated automatically from this. Typically, for maximum sensitivity, at 100 μm pixel size, the speed of the stage is set at 100 μm/s, giving a scan rate of 1 s per pixel. However using TOF mass spectrometers, the scan rate can be up to 30 scans per second—for the same pixel size of 100 μm the stage will therefore move at 3 mm/s.
5. MS parameters such as the type of experiment (MS, MS/MS, in TOF mode only or with ion mobility separation (IMS)), mass range, polarity, collision energy are defined also in HDI. Processing parameters can also be set prior to the acquisition. Save the experiment parameters and create the worksheet to be loaded into MassLynx.
6. Start the acquisition using MassLynx, with the optional automatic processing step (*see* **Note 13**).
7. Load the processed DESI imaging data into HDI for visualization of the DESI imaging experiment (*see* Fig. 5 and for DESI images *see* Fig. 6).

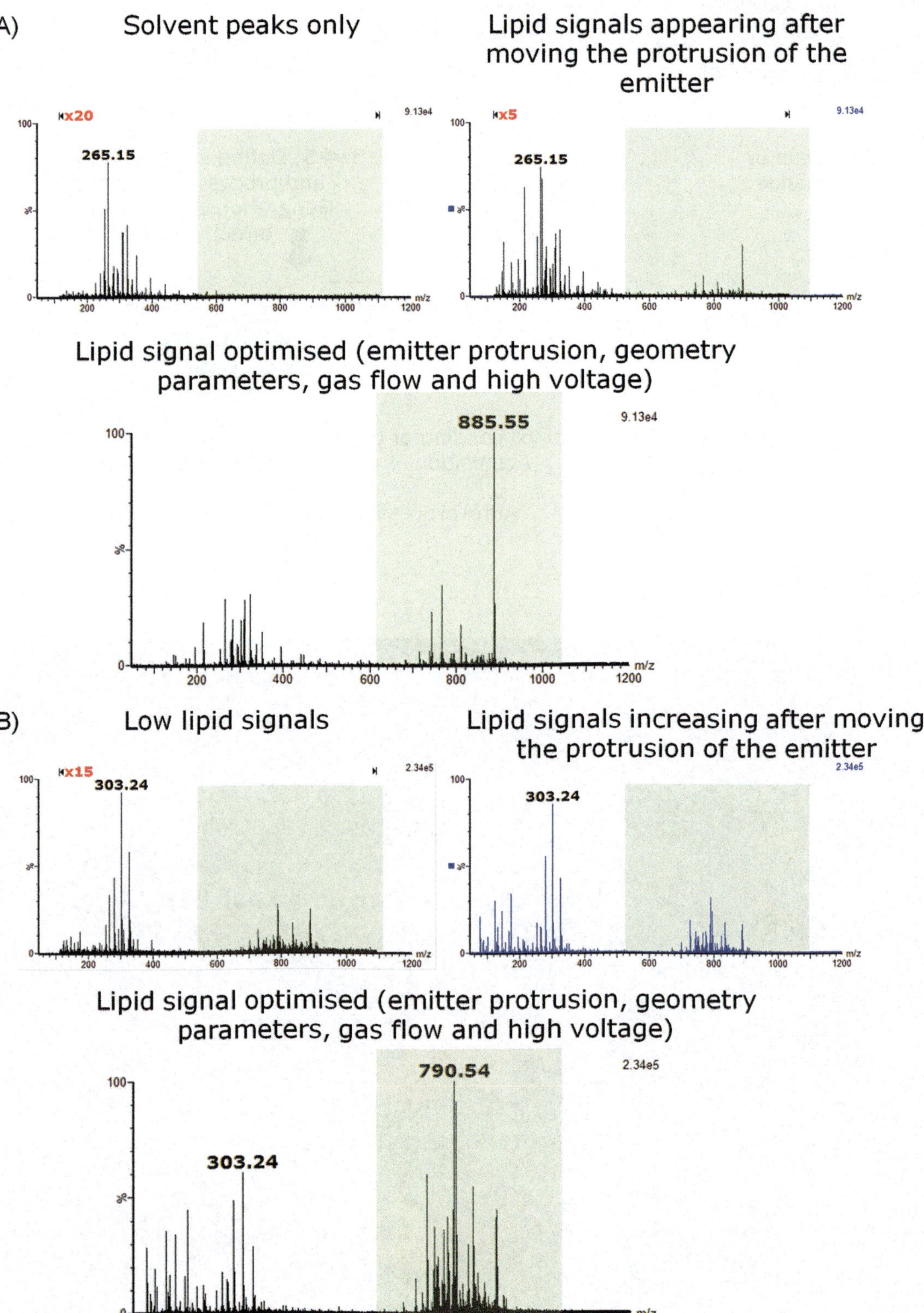

Fig. 4 (**a**) MS spectra for DESI optimization of lipid signals on *porcine liver* tissue section; (**b**) MS spectra for DESI optimization of lipid signals on mouse brain tissue section

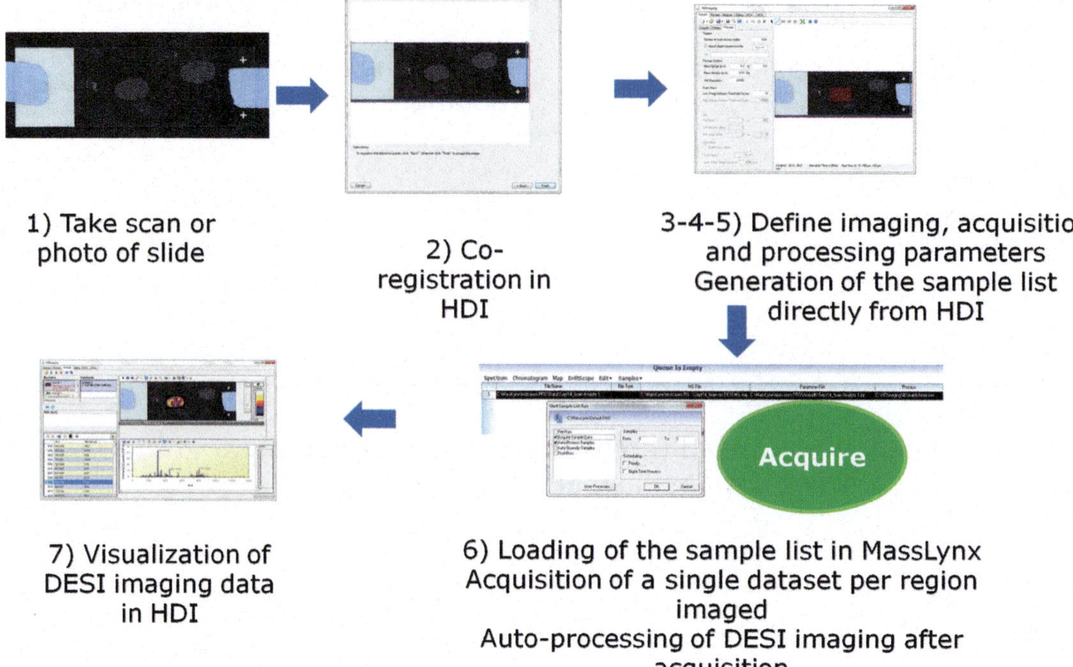

Fig. 5 DESI IMS workflow

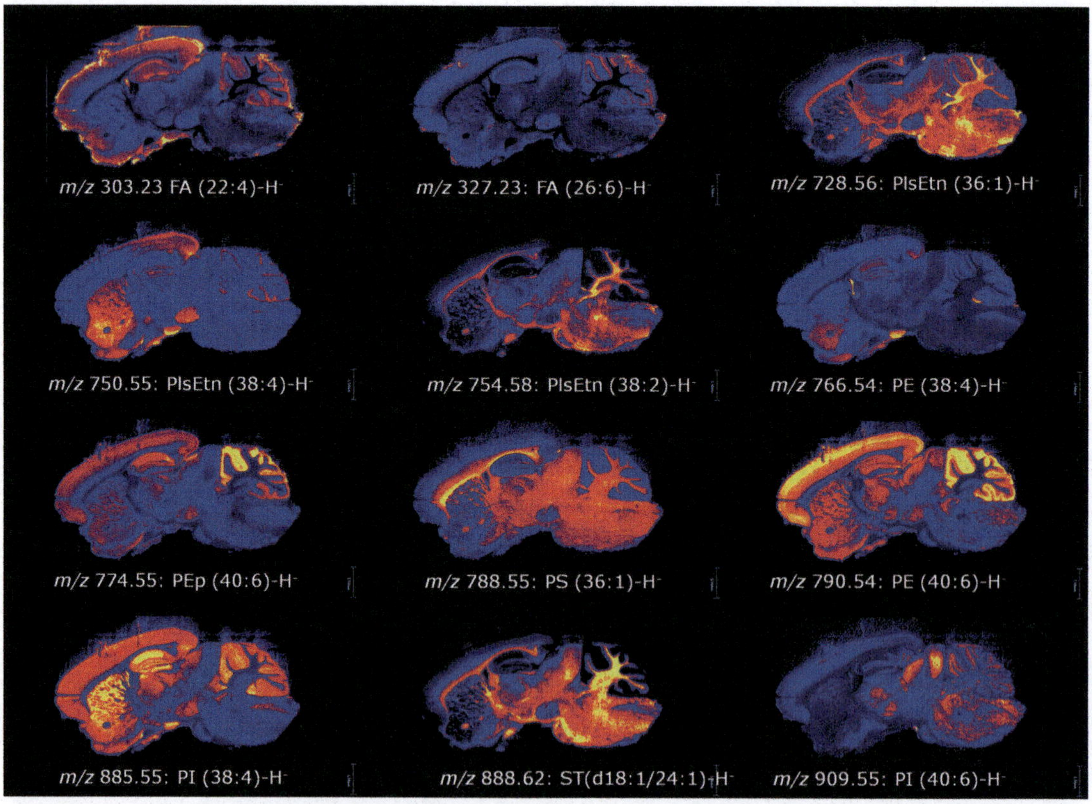

m/z 303.23 FA (22:4)-H⁻
m/z 327.23: FA (26:6)-H⁻
m/z 728.56: PlsEtn (36:1)-H⁻
m/z 750.55: PlsEtn (38:4)-H⁻
m/z 754.58: PlsEtn (38:2)-H⁻
m/z 766.54: PE (38:4)-H⁻
m/z 774.55: PEp (40:6)-H⁻
m/z 788.55: PS (36:1)-H⁻
m/z 790.54: PE (40:6)-H⁻
m/z 885.55: PI (38:4)-H⁻
m/z 888.62: ST(d18:1/24:1)-H⁻
m/z 909.55: PI (40:6)-H⁻

Fig. 6 Examples of ion images generated from a DESI IMS experiment on a mouse brain at 40 μm pixel size in negative mode

4 Notes

1. The emitter can be purchased precut at 40 mm from Waters Corporation or needs to be cut when purchased from New Objective using a diamond blade capillary cutter such as the Shortix cutter from Supelco.
2. Make sure the emitter is fixed in position using plastic tweezers. If the emitter is not tightly held, it will be in danger of coming loose during operation and becoming damaged.
3. Hold the main body horizontally to allow for the spray cone to be introduced in a way that the thread finds its position and gets centered before the emitter gets close to the inner surfaces.
4. To avoid damage to the emitter tip, it is recommended that all work is done with the stage lowered and the arm moved outward.
5. The fitting for the solvent capillary does not require a sleeve and is for 360 μm o.d. fused silica, alternative fittings will be required for other solvent capillary outer diameters.
6. The high desorption/ionization efficiency of these ink samples allows for a preliminary check that the system is operational and to obtain suitable starting positions for the optimization on tissue. Due to the very different nature of these surfaces, it should not be assumed that a setup that yields the greatest signal for the ink will be the optimal conditions for the tissue sections under investigation. Following this preliminary procedure unsing ink, reoptimize the set-up using a tissue section, ideally a tissue section of the same origin as those intended for the study. This optimization step need only be carried out in full on first installation or after changing the emitter in the sprayer or if a completely different sample type is to be analyzed.
7. When ready to begin the imaging study, take the glass slide with the tissue section mounted from the −80 °C degrees freezer. Let it defrost at room temperature under nitrogen gas, then optionally in vacuum desiccator for less than 30 min. No other sample preparation is required. Place the sample onto the DESI stage.
8. Solvent composition can vary depending on the analytes of interest and also the surface. For mammalian tissue sections, 95–98% aqueous methanol solvent composition is a good starting point for the analysis of metabolites and lipids for both negative and positive ion modes. 0.1% Formic acid can be added to the solvent for positive ion mode analysis.
9. Solvent delivery needs to be stable, otherwise fluctuation and signal instability will result that will subsequently be reflected in the ion images. Refilling the syringe in the middle of a DESI

imaging experiment will affect the spray and therefore the quality of the ion images. With the mass spectrometer internal fluidics, the syringe volume options are 250 or 500 μL. Ideally using a 1 mL syringe allows a DESI imaging experiment to last for 11 h at 1.5 μL/min flow rate.

10. To evaluate whether damage has occurred to the emitter tip, check it under a microscope if possible, in particularly if the emitter has been knocked. If there is a leak and blockage in the emitter or downstream, check that solvent is arriving at the end of capillary prior to the sprayer and then at the end of the emitter.
11. To change the protrusion of the emitter, lift the lid of the DESI cover. When the lid of the DESI cover is in up position, the high voltage of the mass spectrometer is being turned off. By closing the lid, the high voltage is turned back on.
12. The fiducial points for coregistration of the optical image are the four glass slide corners, so they need to be present in the optical image.
13. Processing parameters can be defined before the acquisition and therefore the processing step is automatically performed following the acquisition.

References

1. Takats Z, Wiseman JM, Gologan B, Cooks RG (2004) Mass spectrometry sampling under ambient conditions with desorption electrospray ionization. Science 306(5695):471–473
2. Friia M, Legros V, Tortajada J, Buchmann W (2012) Desorption electrospray ionization - orbitrap mass spectrometry of synthetic polymers and copolymers. J Mass Spectrom 47:1023–1033
3. Chen H, Talaty NN, Takats Z, Cooks RG (2005) Desorption Electrospray Ionization Mass Spectrometry for High-Throughput Analysis of Pharmaceutical Samples in the Ambient Environment. Anal Chem 77:6915–6927
4. Manicke NE, Nefliu M, Wu C, Woods JW, Reiser V, Hendrickson RC, Cooks RG (2009) Imaging of Lipids in Atheroma by Desorption Electrospray Ionization Mass Spectrometry. Anal Chem 81:8702–8707
5. Bereman MS, Williams TI, Muddiman DC (2007) Carbohydrate Analysis by Desorption Electrospray Ionization Fourier Transform Ion Cyclotron Resonance Mass Spectrometry. Anal Chem 79:8812–8815
6. Takats Z, Wiseman JM, Ifa DR, Cooks RG (2008) Desorption Electrospray Ionization (DESI) Analysis of Tryptic Digests/Peptides. CSH Protoc prot4993. http://cshprotocols.cshlp.org/content/2008/4/pdb.prot4993.full.pdf+html
7. Takats Z, Wiseman JM, Ifa DR, Cooks RG (2008) Desorption Electrospray Ionization (DESI) Analysis of Intact Proteins/Oligopeptides. CSH Protoc prot4992. http://cshprotocols.cshlp.org/content/2008/4/pdb.prot4992.short
8. Girod M, Shi JX, Cheng YZ, Cooks RG (2010) Desorption Electrospray Ionization Imaging Mass Spectrometry of Lipids in Rat Spinal Cord. J Am Soc Mass Spectrom 21:1177–1189
9. Dill AL, Ifa DR, Manicke NE, Costa AB, Ramos-Vara JA, Knapp DW, Cooks RG (2009) Lipid Profiles of Canine Invasive Transitional Cell Carcinoma of the Urinary Bladder and Adjacent Normal Tissue by Desorption Electrospray Ionization Imaging Mass Spectrometry. Anal Chem 81:8758–8764
10. Guenther S, Muirhead LJ, Speller AVM, Golf O, Strittmatter N, Ramakrishnan R, Goldin RD, Jones EA, Veselkov K, Darzi A, Takats Z (2015) Spatially Resolved Metabolic Phenotyping of Breast Cancer by

Desorption Electrospray Ionization Mass Spectrometry. Cancer Res 75:1828–1837

11. Kertesz V, Van Berkel GJ (2008) Improved imaging resolution in desorption electrospray ionization mass spectrometry. Rapid Commun Mass Spectrom 22:2639–2644
12. Gurdak E, Green FM, Rakowska PD, Seah MP, Salter TL, Gilmore IS (2014) VAMAS Interlaboratory Study for Desorption Electrospray Ionization Mass Spectrometry (DESI MS) Intensity Repeatability and Constancy. Anal Chem 86:9603–9611
13. Schwartz SA, Reyzer ML, Caprioli MR (2003) Direct tissue analysis using matrix-assisted laser desorption/ionization mass spectrometry: practical aspects of sample preparation. J Mass Spectrom 38:699–708
14. Badu-Tawiah A, Bland C, Campbell DI, Cooks RG (2010) Non-Aqueous Spray Solvents and Solubility Effects in Desorption Electrospray Ionization. J Am Soc Mass Spectrom 21:572–579
15. Green FM, Salter TL, Gilmore IS, Stokes P, O'Connor G (2010) The effect of electrospray solvent composition on desorption electrospray ionisation (DESI) efficiency and spatial resolution. Analyst 135:731–737

Chapter 8

Peptide Imaging: Maximizing Peptide Yield, Optimization of the "Peptide Mass Fingerprint"

Ekta Patel

Abstract

In Matrix Assisted Laser Desorption Ionization Mass Spectrometry Imaging (MALDI-MSI), identification of observed proteins remains a challenge primarily due to the drop in sensitivity of time-of-flight (TOF) mass spectrometers beyond the mass range of 25–35 kDa (Francese et al. High Throughput Screen 12: 156–174, 2009) and inadequate mass resolving power at those molecular weights. To counteract this, a bottom-up proteomic approach is often employed. Proteolysis digests proteins into smaller peptide fragments, typically between 500 and 3000 Da which are easier to detect with a higher mass accuracy. Here, we describe the addition of a MS-compatible detergent to the endopeptidase trypsin, to enhance in situ proteolysis efficiency and peptide intensity.

Key words MALDI, In situ, Proteolysis, Detergents

1 Introduction

Enzymatic digestion is conventionally carried out in solution. However, in order to maintain spatial integrity and visualize protein localization, in situ digestion must be carried out. Deposition of the serine protease trypsin has been reported in literature in different ways. Sprayers produce a homogeneous coating that is necessary for successful imaging experiments [1, 2, 3]; however, other methods include robotic spotters [4, 5]. Efficient peptide extraction may be difficult to achieve at times due to the hydrophobic nature of some proteins. Proteins exist within cells as tightly associated complexes or are inserted into lipid bilayers [6]. MS-compatible detergents have been previously used to enhance in gel [7] and in solution [6] digestion of hydrophobic proteins, such as membrane proteins. Membrane proteins can be proteolytically resistant to digestion because of inaccessible cleavage sites. As a result the limited number of peptides produced can affect protein identification; the amphiphilic nature of a detergent improves solubilization by unfolding the protein to make it more accessible to enzymatic cleavages.

Laura M. Cole (ed.), *Imaging Mass Spectrometry: Methods and Protocols*, Methods in Molecular Biology, vol. 1618,
DOI 10.1007/978-1-4939-7051-3_8, © Springer Science+Business Media LLC 2017

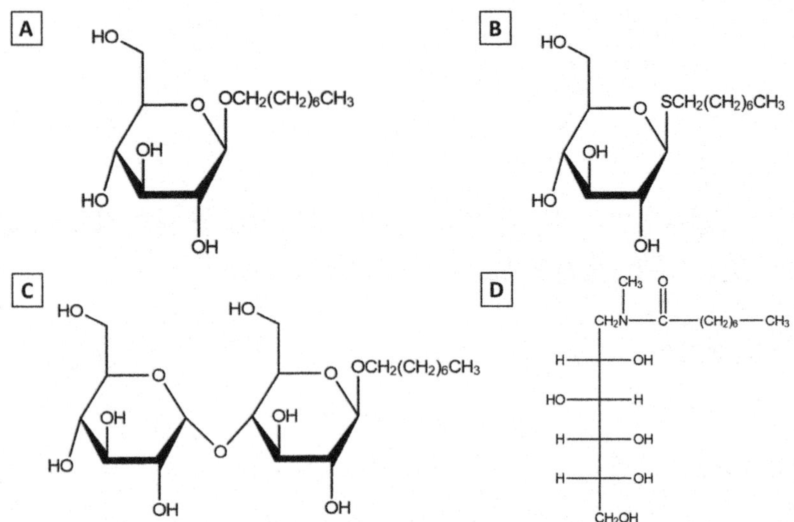

Fig. 1 Chemical structures of nonionic surfactants; (**a**) n-Octyl ß-D-glucopyranoside (OcGlu), (**b**) N-Octyl 1-thio-ß-D-glucopyranoside (OcThio), (**c**) N-Decyl ß-D-maltoside (DDM), (**d**) N-Octanoyl-N-methylglucamin (MEGA-8)

Signal suppression may be an issue when incorporating certain detergents to digest protocols. For example, the ionic detergent sodium dodecyl sulphate (SDS) causes a decrease in the signal to noise ratio and contributes to signal suppression, thus interfering with peptide identification. In past investigations, a comparison between nonionic surfactants and anionic surfactant blends was employed to further improve protein enzymatic digestion from a range of substrates including ungroomed fingermarks and fresh frozen tissue sections [8]. This study involved an investigation into the use of nonionic and anionic surfactants, namely *n*-Octyl ß-D-glucopyranoside (OcGlu), *n*-Octyl 1-thio-ß-D-glucopyranoside (OcThio), *n*-Decyl ß-D-maltoside (DDM), and *N*-Octanoyl-*N*-methylglucamin (MEGA-8) (Fig. 1a–d), to further improve in situ enzymatic digestion [8]. RapiGest SF (Waters Corporation, Manchester, UK) has also been widely reported for its use in solution without interfering with MS analysis [9, 10], however we successfully reported the addition of the surfactant to in situ experiments.

Concentrations of OcGlu higher than 0.1% (up to 2%) were shown to be compatible with MALDI-MS-based analyses and although this detergent is the most effective for efficient proteolysis, others could be employed with even higher efficiencies and reproducibility such as the nonionic surfactant MEGA-8 [8]. The following protocol provided incorporates the addition of a detergent to trypsin for MALDI-MSI experiments.

2 Materials

All solutions were prepared using ultrapure water (18 MΩ-cm at 25 °C) and analytical grade reagents with a purity of ≥99.5%. Reagents should be disposed of in compliance with health and safety regulations and in accordance with COSHH guidelines.

2.1 Materials for Mass Spectrometry Sample Preparation and Acquisition

1. Ethyl alcohol, 200 proof, ACS reagent ≥99.5%.
2. Ammonium bicarbonate, ≥99.5%.
3. Trypsin Gold, Mass Spectrometry Grade.
4. Octanoyl-*N*-Methylglucamide (MEGA-8), Anagrade.
5. Octyl-α/β-glucoside 10 mM solution (OcGlu).
6. α-Cyano-4-hydroxycinnamic acid (CHCA), ≥98% powder.
7. Aniline, ACS reagent ≥99.5%.
8. Acetonitrile, for HPLC, ≥99.9%.
9. Methanol, for HPLC, ≥99.9%.
10. Chloroform ≥99.5%.
11. Trifluoroacetic acid, 99%.
12. Phosphorus red, 99%.
13. Aluminium TLC sheets. SILG/UV254 0.20 mm thickness.
14. Extra adhesive precleaned micro slides.

2.2 Reagents: Working Composition

1. Ethanol solutions: Dilute 100% ethanol with ultrapure water to 90% and 70% (vol/vol).
2. Ammonium bicarbonate 50 mM: Dissolve 0.2 g ammonium bicarbonate in 50 ml ultrapure water. Check and adjust pH to 8.0. Solution can be stored for 1 month.
3. Trypsin solution 20 μg ml^{-1}: Add 1 ml ammonium bicarbonate (pH 8.0) directly to the trypsin 100 μg vial to give a 100 μg ml^{-1} stock solution. Agitate to ensure lyophilized trypsin is reconstituted. For the 20 μg ml^{-1} working solution; pipette 200 μl of the stock solution into a sterile 1.5 ml snap top Eppendorf and add 800 μl ammonium bicarbonate (pH 8.0).
4. MEGA-8 detergent solution 10 mM: Add 32.142 mg of MEGA-8 to 10 ml ultrapure water. Gently swirl to dissolve taking care not to agitate vigorously. Detergent solution can be stored at +4 °C for 3 months.
5. Trypsin solution 20 μg ml^{-1} containing detergent 0.5%: Add 5 μl detergent solution (10 mM) to 995 μl trypsin solution 20 μg ml^{-1}.

6. MALDI matrix solution: Prepare 5 mg ml^{-1} CHCA in 50% (vol/vol) acetonitrile and 0.5% (vol/vol) trifluroacetic acid in ultrapure water. Add equimolar amounts of aniline to the CHCA (i.e., 1 ml of 5 mg ml^{-1} CHCA solution contained 2.4 μl aniline). Sonicate in a sonic water bath for 5–10 minutes until matrix crystals have fully dissolved. Matrix containing aniline should be prepared fresh on the day of the experiment and not stored.
7. Humidity chamber solution: 50% MeOH in water (vol/vol).
8. Phosphorus red calibrant: 10 mg ml^{-1} in 100% acetone. Vortex for 10–15 seconds. N.B. phosphorus red may not dissolve entirely.

3 Methods

3.1 In Situ Digestion of Fingermarks

1. Ungroomed fingermarks were prepared and deposited onto aluminum sheets (TLC sheets were pretreated to remove the stationary phase (*see* **Note 1**)).
2. Seven layers of trypsin were deposited at a 1.5 μl/minute flow rate and a nitrogen pressure of 3 bar using the SunCollect pneumatic sprayer (SunChrom GmbH, Freidrichsdorf, Germany) (*see* **Note 2**).
3. Samples were incubated in a parafilm-covered coplin jar containing the humidity chamber solution; 50% MeOH (Fig. 2), for 3 hours at 37 °C.

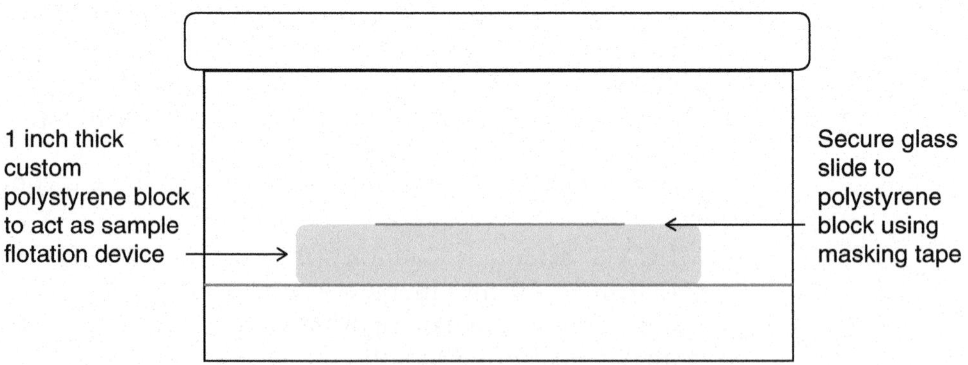

Fig. 2 Humidity chamber setup using a glass coplin jar containing 50:50 MeOH: water and a polystyrene floatation device to hold the sample. Once the lid is placed on the jar carefully wrap the seal with parafilm to allow the humidity to remain within the chamber

3.2 Cryo-conservation of Tissue

1. Arrange the excised organ/biopsy into the correct orientation onto a regular weighing boat.
2. While wearing appropriate PPE, float the weighing boat containing the tissue on liquid nitrogen (*see* **Note 3**).
3. Take ~50 μl of 2% CMC and attach the frozen tissue to the cork circle, ensuring the tissue is level and in the correct orientation.
4. Place into the −20 °C freezer briefly to ensure that the CMC has frozen.
5. Mount the cork circle with tissue to a cryostat chuck. This can be done with ~50–100 μl 2% CMC solution.

3.3 Sectioning and In Situ Digestion of Fresh Frozen Tissue Sections

1. Fresh frozen rat brain tissue sections were cut to 10 μm thickness using a Leica CM3050 cryostat chilled to −20 °C.
2. Sections were thaw-mounted onto Surgipath X-tra adhesive glass slides and stored in glass slide holders at −80 °C (*see* **Note 4**).
3. Before digestion with trypsin, tissue sections were pretreated with wash steps of 70% and 90% ethanol (1 minute each) and chloroform (10 seconds) (*see* **Note 5**).
4. 20 μg ml^{-1} trypsin solution containing 0.5% 10 mM OcGlu or 10 mM MEGA-8 was sprayed onto the tissue sections using the SunCollect automated sprayer (seven layers at a flow rate of 1.5 μl/minute and a nitrogen pressure of 3 bar).
5. Tissue sections were incubated in a parafilm-covered coplin jar containing the humidity chamber solution; 50% MeOH (Fig. 1), for 3 hours at 37 °C.

3.4 Matrix Deposition

1. Five layers of 5 mg ml^{-1} CHCA containing aniline were sprayed onto the samples previously digested with trypsin, using the SunCollect autosprayer (SunChrom GmbH) at a flow rate of 1.5 μl/minute and a nitrogen pressure of 2.5 bar (*see* **Note 6**).
2. Ideally, samples should be analyzed on the same day as spraying matrix. If overnight storage is required due to unforeseen circumstances, keep the sample in a sealed petri dish at +4 °C.

3.5 Instrumentation and Data Acquisition

1. Spot 0.5 μl phosphorus red calibrant onto a clean MALDI target plate. Calibrate the SYNAPT™ G2 HDMS mass spectrometer (Waters Corporation, Manchester, UK) using the phosphorus red spot over 600–2800 Da mass range.
2. Scan the sample sprayed with matrix on a traditional flat-bed scanner and save the image as a .jpeg.
3. Following the dialog box instructions for a *"new image pattern,"* the .jpeg was uploaded within the high definition imaging (HDI) software to allow coregistration and image setup (*see* **Note 7**).

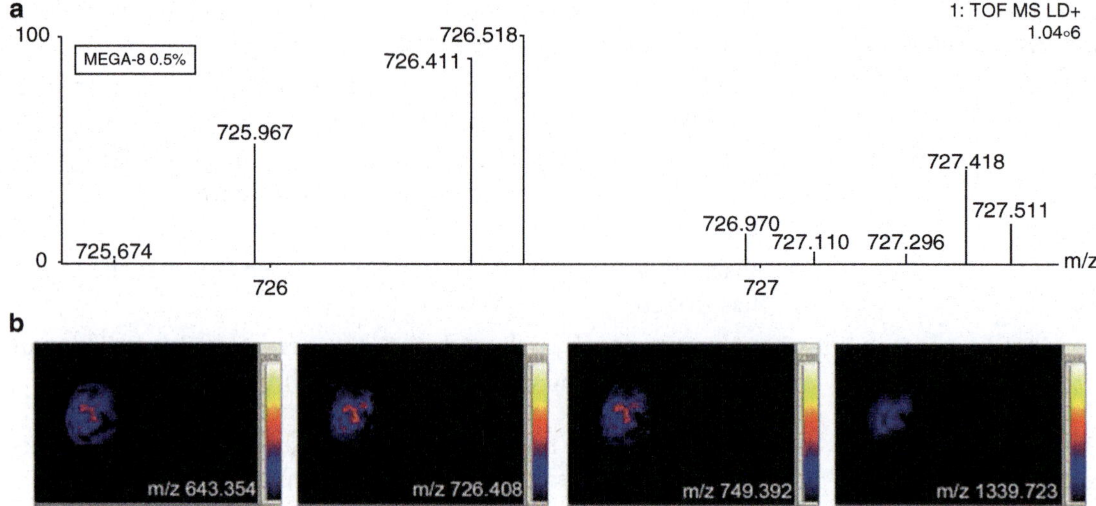

Fig. 3 Peptide profile and images from the digestion of rat brain containing MEGA-8. MALDI MS spectrum shows a very narrow m/z region of the in situ digestion using 20 μg/ml trypsin solution containing MEGA-8 at 0.5%. (**a**) displays the presence of the MBP peptide at m/z 726.4; (**b**) demonstrates the opportunity for chemical mapping through MALDI MSI using 0.5% MEGA-8 as a detergent

4. Once the sample was loaded into the instrument, MALDI-IMS-MS images were acquired in positive ion mode from 600 to 2800 at a mass resolution of 10,000; full width at half maximum (FWHM).
5. The instrument was operated with a 1 KHz Nd:YAG laser and the laser energy was set to 220 arbitrary units on the instrument.
6. Image acquisition was performed at a spatial resolution of 100 μm × 100 μm for the imaging of fingermarks and tissue (Fig. 3).

3.6 Data Analysis

1. Mass spectrometry imaging data were processed using the HDI software (Waters Corporation, Manchester, UK).
2. It is also possible to use open-source software such as Biomap (Novartis) (*see* **Note 8**).
3. HDI software has a number of features including normalization, color gradients, and the option to create overlays of different ions. N.B. Please see the user manual for an extensive guide.

4 Notes

1. Carry out the following procedure in a fume hood and with appropriate PPE. To remove the white silica from TLC plates, first cover the silica side with acetone using a laboratory wash

bottle. Allow the acetone to soak into the silica and top up with more acetone until damp. Using a laboratory hand towel, buff off the silica working from one edge until all of the silica is removed, use more acetone if the silica dries out. N.B. Do not apply excess pressure onto the aluminum plate as this may result in dents. Once the silica is removed use a clean piece of hand towel and wipe away any residues with acetone. Ensure both sides are perfectly clear of silica. Using a scalpel and ruler, carefully score the aluminum plate to the dimensions of a glass slide (2.5 cm width). After scoring gently bend the TLC plate back and forth until each strip breaks off. Score these strips into 7.5 cm length slides and gently break apart. Store in a clean petri dish.

2. To avoid risks of carry over from a previous user, ensure that the capillaries have been flushed clean. When spraying trypsin, first flush the Sun Collect with 100% ACN at a flow rate of 2 μl/minute for 15 minutes. Then flush the capillaries with 50 mM ammonium bicarbonate at the same flow rate for another 15 minutes, prior to spraying trypsin solution. This ensures that any traces of trypsin or matrix from the previous user are removed. After trypsin has been applied repeat the flushing procedure. First, flush through with deionized water at a flow rate of 2 μl/minute for 15 minutes, then with 100% ACN for 15 minutes.
3. Floating the tissue on liquid nitrogen rather than submerging minimizes the chances of freeze fracturing. This allows the tissue to freeze evenly throughout without damaging. At this stage, the tissue can be covered in parafilm and stored in an air-tight container if required at a later date.
4. Thaw mounting: once a section has been cut using the cryostat the tissue section is attached to the glass slide electrostatically; holding the slide close over the section allows it to attach to the glass slide. Rubbing the underside with a gloved finger allows the section to adhere to the slide.
5. Place the 70% and 90% EtOH in a 50 ml falcon tube if one slide is being analyzed at a time. If more than one slide is to be washed at once use a multi slide rack in a glass coplin jar. Ensure that the chloroform is always placed in a glass vessel.
6. Ensure the Sun Collect has been flushed through with 100% ACN at a flow rate of 2 μl/minute for 15 minutes before matrix is flushed through. This allows any matrix crystals blocking capillaries to be dissolved.
7. The HDI software includes automated image processing. This processing feature can be selected during image setup.
8. Images acquired using the SYNAPT™ can be converted using ImageConverter to be viewed on open-source software such as Biomap (Novartis).

Acknowledgments

The author gratefully acknowledges Dr Simona Francese for her support and guidance, and GlaxoSmithKline for the BBSRC Industrial Case Award funding.

References

1. Francese S, Dani FR, Traldi P, Mastrobuoni G, Pieraccini G, Moneti G (2009) MALDI mass spectrometry imaging, from its origins up to today: the state of the art. Comb Chem High Throughput Screen 12:156–174
2. Schober Y, Guenther S, Spengler B, Römpp A (2012) High-resolution matrix-assisted laser desorption/ionization imaging of tryptic peptides from tissue. Rapid Commun Mass Spectrom 26:1141–1146
3. Cole LM, Djidja MC, Bluff J, Claude E, Carolan VA, Paley M, Tozer GM, Clench MR (2011) Investigation of protein induction in tumor vascular targeted strategies by MALDI MSI. Methods 54:442–453
4. Stauber J, MacAleese L, Franck J, Claude E, Snel M, Kaletas BK, Wiel IMVD, Wisztorski M, Fournier I, Heeren RMA (2010) On-tissue protein identification and imaging by MALDI-ion mobility mass spectrom- etry. J Am Soc Mass Spectrom 21:338–347
5. Groseclose MR, Andersson M, Hardesty WM, Caprioli RM (2007) Iden- tification of proteins directly from tissue: in situ tryptic digestions coupled with imaging mass spectrometry. J Mass Spectrom 42:254–262
6. Chen EI, Cociorva D, Norris JL, Yates JR III (2007) Optimization of mass spectrometry-compatible surfactants for shotgun proteomics. J Proteome Res 6:2529–2538
7. Sutton CW, Pemberton KS, Cottrell JS, Corbett JM, Wheeler CH, Dunn MJ, Pappin DJ (1995) Identification of myocardial proteins from two- dimensional gels by peptide mass fingerprinting. Electrophoresis 16:308–316
8. Patel E, Clench MR, West A, Marshall PS et al (2015) Alternative surfactants for improved efficiency of in situ tryptic proteolysis of fingermarks. J Am Soc Mass Spectrom 26:862–872
9. Huang HZ, Nichols A, Liu D (2009) Direct identification and quantification of aspartyl succinimide in an IgG2 mAb by RapiGest assisted digestion. Anal Chem 81:1686–1692
10. Yu YQ, Gilar M, Lee PJ, Bouvier ES, Gebler JC (2003) Enzyme-friend- ly, mass spectrometry-compatible surfactant for in-solution enzymatic di- gestion of proteins. Anal Chem 75:6023–6028

Chapter 9

MALDI-MS Imaging in the Study of Glomerulonephritis

Andrew Smith, Manuel Galli, Vincenzo L'Imperio, Fabio Pagni, and Fulvio Magni

Abstract

Glomerulonephritis (GNs) are one of the most frequent causes of chronic kidney disease (CKD), a renal condition that often leads to end-stage renal failure, and a careful assessment of these diseases is essential for prognostic and therapeutic purposes. The application of MALDI-MSI directly on bioptic renal tissue represents a new stimulating perspective and facilitates the detection of specific proteomic indicators that are directly correlated with the pathological alterations that occur within the glomeruli during the development of glomerulonephritis. Here, we describe the standard workflow for the MALDI-MSI analysis of clinical fresh-frozen and FFPE renal biopsies and highlight how the obtained molecular information, when combined with histology, can be used to detect specific protein markers of GNs.

Key words Kidney, Glomerulonephritis, MALDI-imaging, Proteomics

1 Introduction

Glomerulonephritis (GNs) are one of the most frequent causes of chronic kidney disease (CKD), a renal condition that often leads to end-stage renal failure, and a careful assessment of these diseases is essential for prognostic and therapeutic purposes [1]. The renal biopsy still plays a crucial role for the distinction among the various types of glomerular diseases [2, 3]. However, it is considered an invasive procedure affected by a certain rate of complications [4, 5], so in recent decades, many efforts were made to identify biomarkers able to discriminate among CKDs on less invasive and easy-to-collect samples such as urine and serum. In this context, the employment of various proteomics techniques has played a crucial role in determining molecular changes related to disease progression and early pathological glomerular modifications [6, 7]. Finally, the application of MALDI-MSI directly on bioptic renal tissue in order to highlight differences in protein expression among a wide range of glomerular conditions represents a new stimulating perspective [8]. Through the correlation of molecular

Laura M. Cole (ed.), *Imaging Mass Spectrometry: Methods and Protocols*, Methods in Molecular Biology, vol. 1618,
DOI 10.1007/978-1-4939-7051-3_9, © Springer Science+Business Media LLC 2017

with histological information, along with the direct collaboration with nephropathologists, facilitates the detection of specific proteomic indicators that are directly correlated with the pathological alterations that occur within the glomeruli, or tubules, during the development of glomerulonephritis. Given the small amount of tissue made available through the bioptic procedure (renal biopsies are commonly ~10 mm in length and ~3 mm in diameter), correlation with the histological features represents a crucial step in the study of glomerular diseases.

Mainini et al. first applied MALDI-MSI to human renal tissue obtained by biopsy in order to evaluate the feasibility of this modern proteomic approach. Interestingly, it was determined that the glomeruli and tubules of healthy tissue presented similar proteomic profiles. However, in the case of primary glomerulonephritis, glomeruli and tubules presented different protein profiles. Furthermore, altered protein expression compared to controls was evident between different types of primary glomerulonephritis, such as membranous nephropathy (MN) and minimal change disease (MCD). Finally, tubules affected by GN, even without morphological evidence of the disease, showed a different protein profile compared to controls. Thus, it is possible to detect early molecular alterations of the disease that are not accessible by traditional histological methods. This feasibility study highlighted the potential role that MALDI-MSI could play in the detection of diagnostic biomarkers associated with primary glomerulonephritis [9].

This body of work was further expanded upon by Smith et al. in 2015, applying MALDI-MSI to bioptic renal tissue taken from patients with the most frequent glomerular kidney diseases (GKDs); Focal Segmental Glomerulosclerosis (FSGS), IgA Nephropathy (IgAN), and MN. First, this technique was able to generate molecular signatures capable of distinguishing between normal kidney and pathological GN, with specific signals representing potential biomarkers of CKD progression. Furthermore, specific disease-related signatures for FSGS and IgAN were detected. Among the specific FSGS-related signatures, one protein was identified as α-1-antitrypsin and, upon validation with the antibody, was shown to be localized to the podocytes within sclerotic glomeruli. This work showed a promising application of MALDI-MSI in the study of glomerulonephritis, highlighting a number of potential biomarkers of CKD progression [10].

In order to bring such MALDI-MSI information into a clinical setting, these findings should be correlated with information obtained from the urinary proteome or peptidome in order to highlight whether tissue-derived proteins of glomerular diseases can also be detected in this more easily accessible sample type. If successful, such biomarkers could be translated into less-invasive diagnostic or prognostic tools, which is the ultimate goal of many clinical proteomics applications.

Here, an overview to the sample preparation procedures for the MALDI-MSI analysis of bioptic renal tissue that has been obtained through the routine clinical workflow is provided. Although the workflow is similar to what would be required for the analysis of fresh-frozen and formalin-fixed paraffin-embedded (FFPE) tissue, there are specific points to consider related to the embedding material and the size of the morphological features. Furthermore, the histological evaluation of the tissue performed following MALDI-MSI analysis is a crucial step when studying glomerular diseases and should be performed carefully with the assistance of a nephropathologist when possible. It is through this collaboration, along with data elaboration procedures such as virtual microdissection, that the extraction of pathologically significant information from this limited amount of tissue can be obtained.

2 Materials

1. OCT Embedding Medium.
2. Indium Tin Oxide-coated glass slides, or other MALDI targets (instrument compatible).
3. Ethanol Solutions (A, B, C, D: 70, 90, 95, 100%).
4. Trypsin solution: 100 ng/μL sequencing grade modified trypsin in 40 mM NH_4HCO_3.
5. Sinnapinic acid (SA) matrix solution: 10 mg/mL in 60/40 ACN:H_2O w/0.2% TFA.
6. α-CHCA matrix soluton: 10 mg/mL in 50:50 ACN:H_2O w/0.4% TFA.
7. Protein Calibration Standard: Angiotensin II, Angiotensin I, Substance P, Bombesin, ACTH clip 1–17, ACTH clip 18–39, Somatostatin 28.
8. Peptide Calibration Standard: Insulin, Ubiquitin I, Cytochrome C, Myoglobin.
9. Antigen retrieval buffer: 10 mM citric acid prepared at pH 6.0 and preheated to 65 °C.

2.1 Instrumentation

1. Cryostat.
2. Microtome.
3. Tissue Microarrayer.
4. Standard desktop digital scanner.
5. Scanscope CS2 digital scanner.
6. *iMatrix Spray* (for matrix and trypsin deposition).
7. MALDI Mass Spectrometer: UltrafleXtreme.

2.2 Software

1. *FlexControl* 3.4 (Bruker Daltonik GmBH, Bremen, Germany)—for setting the parameters for spectral acquisition.
2. *FlexImaging* 3.0 (Bruker Daltonik GmBH, Bremen, Germany)—for image acquisition and visualization.
3. *SCiLS Lab 2014* (http://scils.de/; Bremen, Germany*)*—for extensive data elaboration.

3 Methods

3.1 MALDI-MSI of Intact Proteins Using Fresh Frozen Tissue

3.1.1 Tissue Sectioning

1. A tissue area if interest is excised from the surgical nephrectomy, placed into a 50 mL falcon tube and immediately frozen in liquid nitrogen, avoiding any mechanical damage to the tissue during the procedure. In the case of small animals, it is also possible to freeze entire organs. The tissue is then stored at −80 °C.
2. At the time of cutting, the tissue is removed from the freezer (in cold conditions) and taken to the cryostat where it is brought to the cutting temperature (−20 °C). The block of tissue is then mounted on an embedding medium (OCT), trimmed and cryo-sectioned into 5–10 μm sections. The temperature at which the cryo-sectioning is performed should be selected based upon the morphological characteristic of the tissue under investigation and the abundance of the material.
3. Before mounting sections for MALDI-MSI analysis, it is possible to mount sections on standard microscope glass slides for staining with H&E in order to confirm that the desired regions of interest are present for analysis.
4. Sections for MALDI-MSI are thaw-mounted onto compatible MALDI targets/ITO coated slides. In cases when ITO glass slides are not used as MALDI targets, it is highly recommended to mount consecutive sections on standard microscope slides in order to align and overlap histological and MALDI-MS images.
5. The tissue proteome is then stabilized in an oven at 85 °C for 30 min.
6. The ITO glasses are then packaged as shown in Fig. 1 and stored in a freezer at −80 °C until the day of MALDI-MSI analysis.

3.1.2 Tissue Preparation for MALDI-MSI

1. Following removal from the freezer, the ITO glass slides with mounted sections are dried via desiccation for 30–60 min (based upon the thickness and dimensions of the section). Vacuum dehydration removes excessive moisture from the surface of the tissue that may occur during the thaw-mounting process, thus reducing the possibility of analyte delocalization.
2. The excess OCT compound is removed from the edges of the tissue by using a swab that has been briefly dipped in a 70% EtOH solution. Be careful when approaching the edges of the

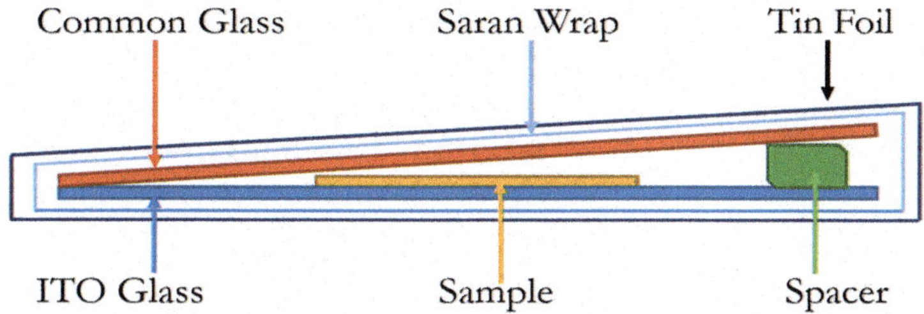

Fig. 1 Sample stocking procedure for ITO glass slides with mounted tissue

tissue, if too much force is used the edges of the tissue may be broken and detached from the ITO glass slide (*see* **Notes 1** and **2** of Sect. 4).

3. The tissue sections are then placed sequentially into petri dishes containing increased concentrations of Ethanol (A, B, C) and gently agitated for 30 s during each wash. Excess Ethanol is then quickly removed from the slide by wiping gently with tissue paper.
4. Teaching points are positioned on the corner of the slide/target prior to scanning. Scanning of the tissue is performed prior to matrix deposition in order to obtain a clear image of the tissue borders. The resolution of the scanned image should be selected based upon the spatial resolution of the planned MALDI-MSI analysis (2400 dpi is usually sufficient for 30 μm spatial resolution MALDI-MSI acquisitions).
5. The SA matrix solution is deposited onto the tissue using the *iMatrix Spray* automated spraying device. An example "dry spray" method involves the following parameters; Spray height: 60 mm, Speed: 180 mm/s, Line distance: 1 mm, Density: 2 μL/cm^2, Break: 60 s and Number of cycles: 5 [11] (*see* **Note 3** of Sect. 4).
6. When conductive ITO glass slides are used, matrix is removed from the edges of the slide using Methanol and tissue paper in order to avoid uneven conductivity of the slide. Matrix is also removed from a small area near to the tissue in order to add protein calibrants.
7. A solution containing a 1:1 ratio between SA and the protein calibrant is prepared. This protein calibrant is then spotted onto a clean area of the ITO slide/target.

3.1.3 MALDI-MSI Analysis

1. The previously positioned teaching points are used for teaching the instrument and to set the *X* and *Y*-axes. The measurement region is then set using *FlexImaging* software and information related to the geometry is sent to the instrument using *FlexControl*. Other parameters, such as the rastering, are configured in *FlexImaging* and selected based upon the dimen-

sion of the histological features to be analyzed and the crystal dimensions. Considering the dimensions of glomeruli, a raster of ≤50 μm is recommended. For each raster position, a mass spectrum is acquired by accumulating approximately 500 shots. The number of shots should be carefully selected by the user depending upon the amount of matrix deposited and the fragility of the tissue.

2. Following calibration using the protein standard, the selected section is analyzed using a Bruker UltrafleXtreme mass spectrometer equipped with a Smartbeam laser (Nd:YAG, 355 nm wavelength, 1–2 kHz frequency) and operated in positive linear mode in the mass range of 3–20 kDa.

3.1.4 Histological Evaluation

1. Following MALDI-MSI analysis (Fig. 2), the matrix can be removed from the tissue by washing sequentially in increasing concentrations of Ethanol (A, B, D) and gently agitated for 30 s during each wash. Excess Ethanol is then quickly removed from the slide by wiping gently with tissue paper.
2. The tissue is then colored with the appropriate histochemical stain(s) (*see* **Note 4** of Sect. 4) in order to highlight the particular morphological features that are desired in the study (*see* Table 1).
3. The stained tissue section is then digitally scanned for the direct overlap of histological and MALDI-MS images.

3.2 MALDI-MSI of Tryptic Peptides from Archived FFPE Tissue

3.2.1 Tissue Sectioning

1. FFPE samples are fixed in formalin immediately following surgical excision. The time of fixation is related to the size and dimension of the tissue, but is most commonly between 24 and 48 h. Following formalin fixation, excess formalin is removed by washing the tissue with water. The tissue is then subjected to an automated fixation procedure involving ethanol dehydration, a xylene bath, and paraffin fixation.
2. FFPE blocks are stored at −20 °C before cutting in order to improve section quality. Sections are cut to a thickness of 5 μm by a microtome at RT.

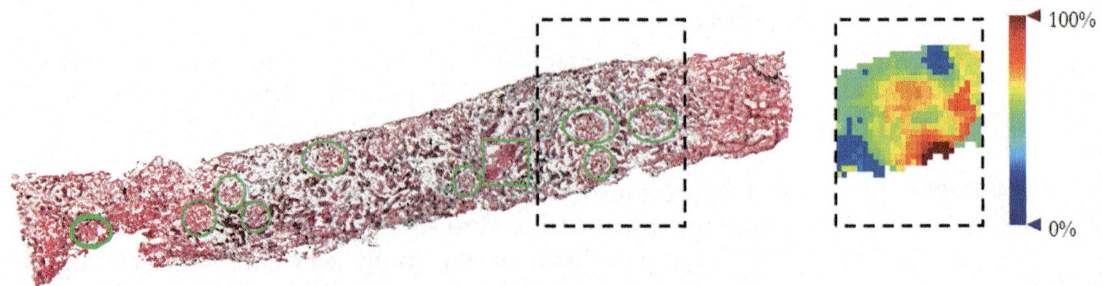

Fig. 2 (*Left*) Example renal biopsy that has been stained using Masson's trichrome, with the pathological glomeruli (*green circle*) and tubules (*green rectangle*) annotated by the nephropathologist. (*Right*) MALDI-MSI representation from an annotated pathological region (*dashed black rectangle*)

Table 1
Routine immunohistochemical stains used in the evaluation of renal biopsies

Immunohistochemical stain	Diagnostic feature(s) to be evaluated
Hematoxylin & Eosin (H & E)	General evaluation, cellular features, presence of inflammation, interstitial oedema, tubular and epithelial damage
Jones methanamine silver (JMS)	Glomerular basement membranes, tubulitis, miotic figures
Masson's trichrome	Extracellular glomerular matrix, tubular basement membranes, fibrosis connective tissue, Bowman' capsule
Periodic acid-Schiff (PAS)	General evaluation of glomeruli, glomerular basement membrane, Bowman' capsule, mesangium, tubular basement membrane

3. FFPE sections are transferred into a milliQ water bath at 37 °C in order to unfold them and are then mounted onto a conductive ITO glass slide. Excess water is removed by leaving the glass slide on a warm plate (at 37 °C) for a few minutes. It is important to monitor the time of exposure to the warm plate as long exposure times can lead to damage of the tissue section and poor adhesion to the slide.

3.2.2 Tissue Preparation for MALDI-MSI Analysis

1. Deparaffinization is performed as follows: Slides are placed in an oven at 65 °C for 1 h and then immersed three times in HPLC grade xylene (three times for 5 min) in order to dewax the tissue. The tissue is then rehydrated by washing in 100% EtOH (twice for 3 min), 70% EtOH (once for 3 min), and HPLC grade H_2O (twice for 2 min).
2. Antigen retrieval is performed in a bath of 10 mM citric acid buffer for 45 min. It is important to note that the temperature should be slowly brought to 97 °C in order to avoid damage to the tissue. After the 30-min period, the bath is cooled to RT.
3. The slides are then washed with HPLC grade H_2O for 2 min and excess water removed by using paper towel. The slide is then further dried under vacuum, via desiccation, for 10 min.
4. Trypsin deposition is performed using the *iMatrix Spray* automated spraying device.

5. It is imperative to monitor and optimize the incubation and drying times during trypsin deposition, based upon the tissue characteristics and dimensions. The incubation phase should be sufficiently long to allow the trypsin to permeate the tissue, while the drying phase should leave the tissue section completely dry. If these phases are not optimized, and the sample becomes too wet, it can lead to analyte delocalization.
6. Following trypsin deposition, the slide is placed in a humidity chamber, containing HPLC grade H_2O, at 45 °C overnight in order for the trypsin digestion to occur.
7. Following trypsin digestion, the slides are then dried under vacuum, via desiccation, for 10 min.
8. Teaching points are positioned on the glass slide and scanned.
9. The α-CHCA matrix solution is then deposited using the *iMatrix Spray* automated spraying device (*see* **step 11** of Subheading 3.1.2).
10. When conductive ITO glass slides are used, matrix is removed from the edges of the slide using Methanol and tissue paper in order to avoid uneven conductivity of the slide. Matrix is also removed from a small area near to the tissue in order to add peptide calibrants.
11. A solution containing a 1:1 ratio between SA and the peptide calibrant is prepared. This protein calibrant is then spotted onto a clean area of the ITO glass slide.

3.2.3 MALDI-MSI Analysis

1. The previously positioned teaching points are used for teaching the instrument and to set the *X* and *Y*-axes. The measurement region is then set using *FlexImaging* software and information related to the geometry is sent to the instrument using *FlexControl*. Other parameters, such as the rastering, are configured in *FlexImaging* and selected based upon the dimension of the histological features to be analyzed and the crystal dimensions. The number of shots should be carefully selected by the user depending upon the amount of matrix deposited and the fragility of the tissue.
2. Following calibration using the peptide standard, the selected section is analyzed using a Bruker UltrafleXtreme mass spectrometer equipped with a Smartbeam laser (Nd:YAG, 355 nm wavelength, 1–2 kHz frequency) and operated in positive reflector mode in the mass range of 700–3500 Da.

3.2.4 Histological Evaluation

1. Following MALDI-MSI analysis, the matrix can be removed from the tissue by washing sequentially in increasing concentrations of Ethanol (A, B, D) and gently agitated for 30 s during each wash. Excess Ethanol is then quickly removed from the slide by wiping gently with tissue paper.

2. The tissue is subsequently colored with the appropriate histochemical stain(s) and digitally scanned for the direct overlap of histological and MALDI-MS images (*see* Table 1).

3.3 Data Elaboration

3.3.1 Data Preprocessing

1. The acquired data can then be visualized in *FlexImaging* for preliminary evaluation and preprocessing. The data can then be imported into *SCiLS Lab 2014* for more intensive processing and data elaboration.
2. A series of preprocessing steps are performed on the loaded spectra: baseline subtraction (TopHat algorithm) and normalization (total ion current algorithm). A series of further steps are performed in order to generate an average (avg.) spectrum representative of the whole measurement region and of specific disease subclasses: peak picking (orthogonal matching pursuit algorithm), peak alignment (to align the detected ions with peak maxima), and spatial denoising [10].
3. Using the coregistered histological image as a guide, virtual microdissection can be performed in order to extract the spectra from only the pathological glomeruli and, thus, obtain glomerular proteomic profiles.
4. Principal Component Analysis (PCA) can also be performed to reduce the high complexity of the data and to monitor spectral similarities between disease subclasses. Finally, Receiver Operative Characteristic (ROC) analysis is performed to discriminate between disease/glomerular classes, with an AUC of >0.8 being required, as an additional criterion to the $p < 0.05$, for a peak to be considered statistically significant.

4 Notes

1. The removal of OCT from the borders of fresh-frozen tissue is a delicate procedure. The OCT is well adhered to the tissue, and thus, if excess force is applied, tissue fragments may be removed along with the OCT. To counter this, try to ensure that the swab remains sufficiently wet in order for the OCT to dissolve somewhat and present a more gel-like state.
2. The swab used for the OCT removal should be both firm and small enough to be manoeuvered very finely around the edges of the tissue.
3. The average glomerulus is approximately 100 μm in diameter, and thus sufficient matrix homogeneity is crucial in order to maximize the amount of proteomic information that can be obtained. Sublimation may also be an appropriate matrix deposition option for use with renal biopsies. However, in our experience, this dry form of matrix deposition does not lead to

sufficient protein extraction given the small amount of protein content present within the tissue. Therefore, the recommendation is to use an automated sprayer although considering that the direct correlation with the histological information is required and spraying parameters should be optimized in order to ensure that analyte delocalization is limited as much as possible.

4. The selection of the most appropriate histological staining should be made based upon the studied GN type and the histological features to be observed. For example, Masson's trichrome is most useful if the disease affects the extracellular glomerular matrix or tubular basement membrane whereas Periodic acid-schiff (PAS) may be more appropriate when observing changes in the glomerular basement membrane or mesangium, which are particularly important features when studying membranous nephropathy (*see* Table 1).

References

1. National Institutes of Health. National Institute of Diabetes and Digestive and Kidney Diseases (2013) United States Renal Data System. USRDS 2013 Annual data report: atlas of chronic kidney disease and end-stage renal disease in the United States. National Institutes of Health. National Institute of Diabetes and Digestive and Kidney Diseases, Bethesda, MD
2. Cameron JS, Hicks J (1997) The introduction of renal biopsy into nephrology from 1901 to 1961: a paradigm of the forming of nephrology by technologyNo title. Am J Nephrol 17:347–358
3. Korbet SM (2002) Percutaneous renal biopsy. Semin Nephrol 22:254–267
4. Walker PD, Cavallo T, Bonsib SM, Society, A. H. C. on R. B. G. of the R. P (2004) Practice guidelines for the renal biopsy. Mod Pathol 17:1555–1563
5. Hergesell O, Felten H, Andrassy K, Kühn K, Ritz E (1998) Safety of ultrasound-guided percutaneous renal biopsy-retrospective analysis of 1090 consecutive cases. Nephrol Dial Transplant 13:975–977
6. Wilkey DW, Merchant ML (2007) Proteomic methods for biomarker discovery in urine. Semin Nephrol 27:584–596
7. Thongboonkerd V (2008) Biomarker discovery in glomerular diseases using urinary proteomics. Proteomics Clin Appl 2:1413–1421
8. L'Imperio, V., Smith, A., Chinello, C., Pagni, F. & Magni, F. (2015) Proteomics and glomerulonephritis: a complementary approach in renal pathology for the identification of chronic kidney disease-related markers Proteomics Clin Appl 10(4):371-383. doi:10.1002/prca.201500075
9. Mainini V et al (2014) MALDI imaging mass spectrometry in glomerulonephritis: feasibility study. Histopathology 64:901–906
10. Smith A, L'Imperio V, De Sio G, Ferrario F, Scalia C, Dell'Antonio G, Pieruzzi F, Pontillo C, Filip S, Markoska K, Granata A, Spasovski G, Jankowski J, Capasso G, Pagni F, Magni F (2016) α-1-antitrypsin detected by MALDI-imaging in the study of glomerulonephritis: its relevance in chronic kidney disease progression. Proteomics 6(11–12):1759–1766. doi:10.1002/pmic.201500411
11. Stoeckli M, Staab D, Wetzel M, Brechbuehl M (2014) iMatrixSpray: a free and open source sample preparation device for mass spectrometric imaging. Chimia (Aarau) 68:146–149

Chapter 10

Hierarchical Cluster Analysis to Aid Diagnostic Image Data Visualization of MS and Other Medical Imaging Modalities

Arul N. Selvan, Laura M. Cole, Lynne Spackman, Sarah Naylor, and Chris Wright

Abstract

Perceiving abnormal regions in the images of different medical modalities plays a crucial role in diagnosis and subsequent treatment planning. In medical images to visually perceive abnormalities' extent and boundaries requires substantial experience. Consequently, manually drawn region of interest (ROI) to outline boundaries of abnormalities suffers from limitations of human perception leading to inter-observer variability. As an alternative to human drawn ROI, it is proposed the use of a computer-based segmentation algorithm to segment digital medical image data.

Hierarchical Clustering-based Segmentation (HCS) process is a generic unsupervised segmentation process that can be used to segment dissimilar regions in digital images. HCS process generates a hierarchy of segmented images by partitioning an image into its constituent regions at hierarchical levels of allowable dissimilarity between its different regions. The hierarchy represents the continuous merging of similar, spatially adjacent, and/or disjoint regions as the allowable threshold value of dissimilarity between regions, for merging, is gradually increased.

This chapter discusses in detail first the implementation of the HCS process, second the implementation details of how the HCS process is used for the presentation of multi-modal imaging data (MALDI and MRI) of a biological sample, third the implementation details of how the process is used as a perception aid for X-ray mammogram readers, and finally the implementation details of how it is used as an interpretation aid for the interpretation of Multi-parametric Magnetic Resonance Imaging (mpMRI) of the Prostate.

Key words Image perception, Image processing, Segmentation, Computer-aided detection, Computer-aided diagnosis, MALDI, MSI, MRI, Mammogram, DCE-MRI, mpMRI

1 Introduction

Tissue abnormality in medical images is usually related to a dissimilar part of an otherwise homogeneous image. The dissimilarity may be subtle or strong depending on the medical modality and the type of abnormal tissue.

Segmentation and delineation of abnormalities in medical images plays an important role in diagnosis and subsequent treatment planning. For example when an abnormality is identified by

Laura M. Cole (ed.), *Imaging Mass Spectrometry: Methods and Protocols*, Methods in Molecular Biology, vol. 1618, DOI 10.1007/978-1-4939-7051-3_10,

a radiologist on a breast screening mammogram then this case is discussed in detail at a multidisciplinary team (MDT) meeting to determine appropriate follow-up. It is important that the lesion extent is appropriately understood and delineated at the MDT in the communication between radiologist and surgeon concerning such abnormal cases which requires some form of region excised. It is well known that disagreements exist concerning the definition of lesion extent and the accurate specification of the location of lesion edges. This issue is important because of the following:

- The potential extent of treated abnormal area can determine what type of subsequent surgical intervention the woman has and
- Surgeons typically excise more breast material surrounding a lesion to ensure that they have fully removed any malignancy.

To visually perceive abnormalities' extent and boundaries requires substantial experience. Consequently, manually drawn region of interest (ROI) to outline boundaries of abnormalities suffers from the following limitations:

- Abnormalities are heterogeneous, hence in manually drawn region of interest (ROI) due to the limitation of human perception information on abnormalities' heterogeneity remains poorly exploited.
- Also manual ROI suffers from inter-observer variability.

As an alternative (to ROI based), it is proposed the use of a segmentation algorithm, for example by Chandarana et al. [1]. In medical images, segmentation of regions of potential abnormalities is a difficult task because of issues such as spatial resolution, poor contrast, ill-defined boundaries, noise, or acquisition artifacts [2].

Segmentation can be thought as a process of grouping visual information, where the details are grouped into objects, objects into classes of objects, etc. Thus, starting from the composite segmentation, the perceptual organization of the image can be represented by a tree of regions, ordered by inclusion. The root of the tree is the entire scene, the leaves are the finest details, and each region represents an object at a certain scale of observation [3].

Traditionally, segmentation algorithms would "binarize" the boundary map by choosing some threshold. There are two problems with thresholding a boundary map [4]:

- The optimal threshold depends on the application, and
- Thresholding a low-level feature like boundaries is likely to be a bad idea for most applications, since it destroys much information.

For these reasons, the segmentation algorithm should operate on a non-thresholded basis. Nevertheless, one needs to threshold the boundary map in order to perceive the boundaries, but this is done at many levels.

Hierarchical Clustering-based Segmentation (HCS) process is an unsupervised segmentation process that generates a hierarchy of segmented images by partitioning an image into its constituent regions at hierarchical levels of allowable dissimilarity between its different regions. The hierarchy represents the continuous merging of similar, spatially adjacent, and/or disjoint regions as the allowable threshold value of dissimilarity between regions, for merging, is gradually increased [5–7].

This chapter discusses in detail the implementation of the HCS process and how the process is applied for the following:

- HCS aided presentation of multi-modal (Magnetic Resonance Imaging—MRI and Matrix-Assisted Laser Desorption/ Ionization—MALDI, Mass Spectrometry Imaging—MSI) imaging data of biological tissue sample.
- HCS as a perception aid for X-ray mammogram readers.
- HCS aided interpretation of Multi-parametric Magnetic Resonance Imaging (mpMRI).

2 Materials

The input to the HCS process is a two-dimensional matrix of numbers. The two-dimensional matrix of numbers can be the representation of the distribution of masses (of peptide and protein) in a biological sample (in the case of MALDI MSI) or the attenuated X-ray energy (in the case of X-ray images) or the nuclear magnetic resonance signal from the hydrogen atoms in an object (in the case of MRI). In the following three sections, the acquisition details of the input data, for each of the three applications of the HCS process, is discussed separately.

2.1 MRI and MALDI-MSI Data of Tissue Sample

The tissue sample was subcutaneously transplanted mouse fibrosarcoma tumors. The MRI sectional images and subsequently MALDI MSI data were acquired as follows [8]:

- The tumor tissue was embedded in gelatin blocks and markers were placed in all four corners of the gelatine blocks in order to aid spatial recognition and registration between modalities (Fig. 1).
- MRI images were acquired using the 0.25 T Esaote GScan. The sample was centrally placed with a dedicated wrist coil and a range of sequences performed FOV (160 × 160). Optimal results were achieved from the T2-weighted Gradient Echo (3NEX) and XBone (4NEX) sequences; 2 mm slices (*see* **Notes 1** and **2**) (Fig. 2).

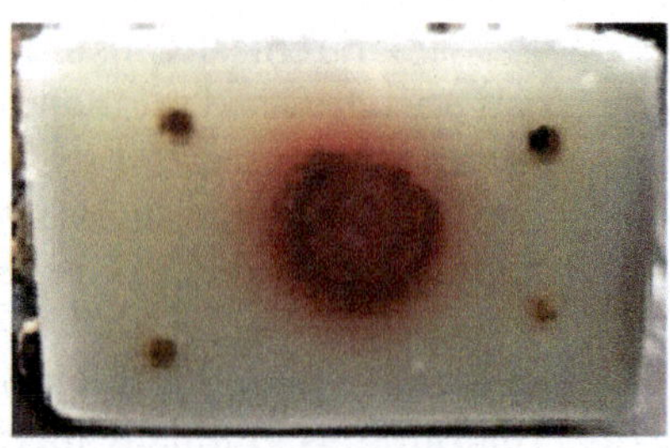

Fig. 1 Tumor tissue embedded in a gelatin block and four markers placed in four corners

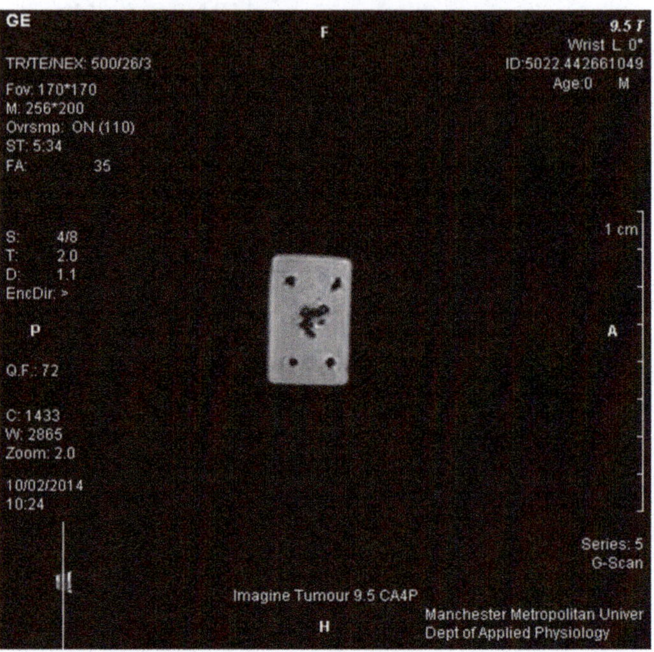

Fig. 2 T2-weighted MRI image of the tissue specimen

- The tissue sample was frozen and then cryosectioned prior to MALDI MSI data capture.
- Peptide mass fingerprints and MALDI Images were performed using the Applied Biosystems Qstar Pulsar i.

2.2 X-Ray Mammogram Image Data

Anonymized X-ray mammogram image data were acquired as part of a screening program using Full Field Digital Mammography (FFDM) system with the pixel spacing of 70 μm (0.070 mm) (*see* **Note 3**). The approximate location and extent of abnormalities were marked by an expert radiologist (Fig. 3).

2.3 mpMRI Image Data

Anonymized MRI image data were acquired as part of diagnostic investigation. The images were acquired with pixel spacing of 1000 μm (1.0 mm) and 1.0 mm spacing between slices. The multi-parametric

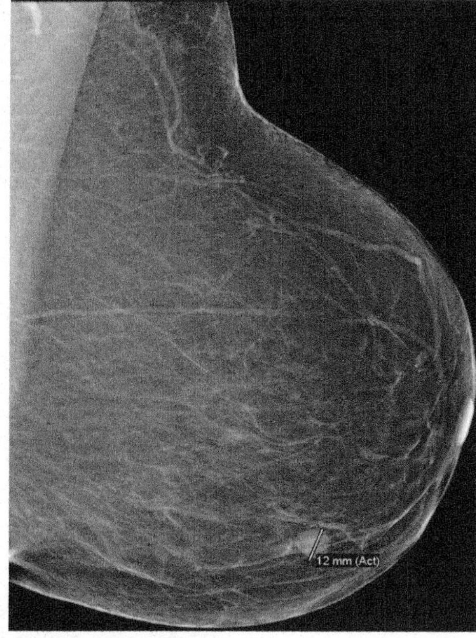

Fig. 3 X-ray mammogram image with the location and size of the abnormality marked by the expert

images were acquired as T2-weighted before injecting intravenous contrast (Fig. 4) and T1-weighted Dynamic Contrast Enhanced (DCE) images. In T2-weighted images, with a small "Field Of View" (FOV), tumour will appear as signal loss (dark).

The DCE MRI images are acquired after intravenously injecting gadolinium chelate contrast (Fig. 5). After injecting intravenous contrast the temporal sections were acquired 16 times at an interval of approximately 10.49 s. Figure 6 lists the Prostate part of the section acquired from time 10:23:51.6275 (SER26) to time 10:26:33.9775 (SER11) (Fig. 6).

3 Methods

In the interpretation of medical images visually perceiving the different component regions in the image is a difficult task, this is due to issues like spatial resolution, poor contrast, ill-defined boundaries, noise, or acquisition artifacts [2]. To aid the visual perception and thus facilitate the interpretation process, the HCS process was adopted to outline and highlight the different regions in an image.

In the following sections, first the HCS process method is outlined. In subsequent sections, it will be discussed in detail how the HCS process was used for the following:

- To visualize the correlating information between the two modalities MRI and MALDI-MSI.

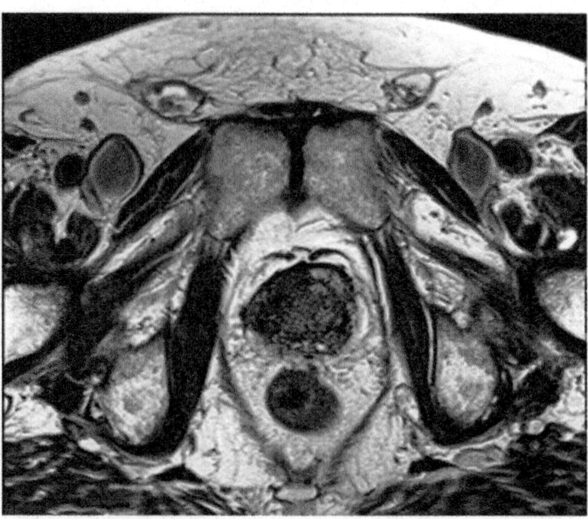

Fig. 4 T2-weighted MRI image section of the prostate

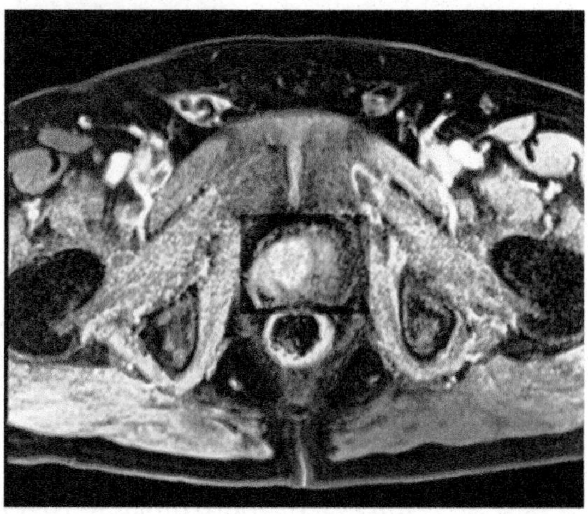

Fig. 5 Dynamic Contrast Enhanced MRI image section of the Prostate post contrast injection

- To visualize the finer details within a potential abnormal region in a X-ray mammogram.
- To visualize multi-parametric MRI image data (T1-weighted DCE-MRI and T2-weighted MRI) and thus correlate information between the two parametric image data.

3.1 Hierarchical Clustering-Based Segmentation (HCS)

Since the early days of computer vision, the hierarchical structure of visual perception has motivated clustering techniques to segmentation [9], where connected regions of the image domain are classified according to an inter-region dissimilarity measure. Hierarchical Clustering-based Segmentation (HCS) [5–7] implements the traditional agglomerative clustering [10], where the regions of an initial partition are iteratively merged and automatically generate a hierarchy of segmented images. The hierarchy of

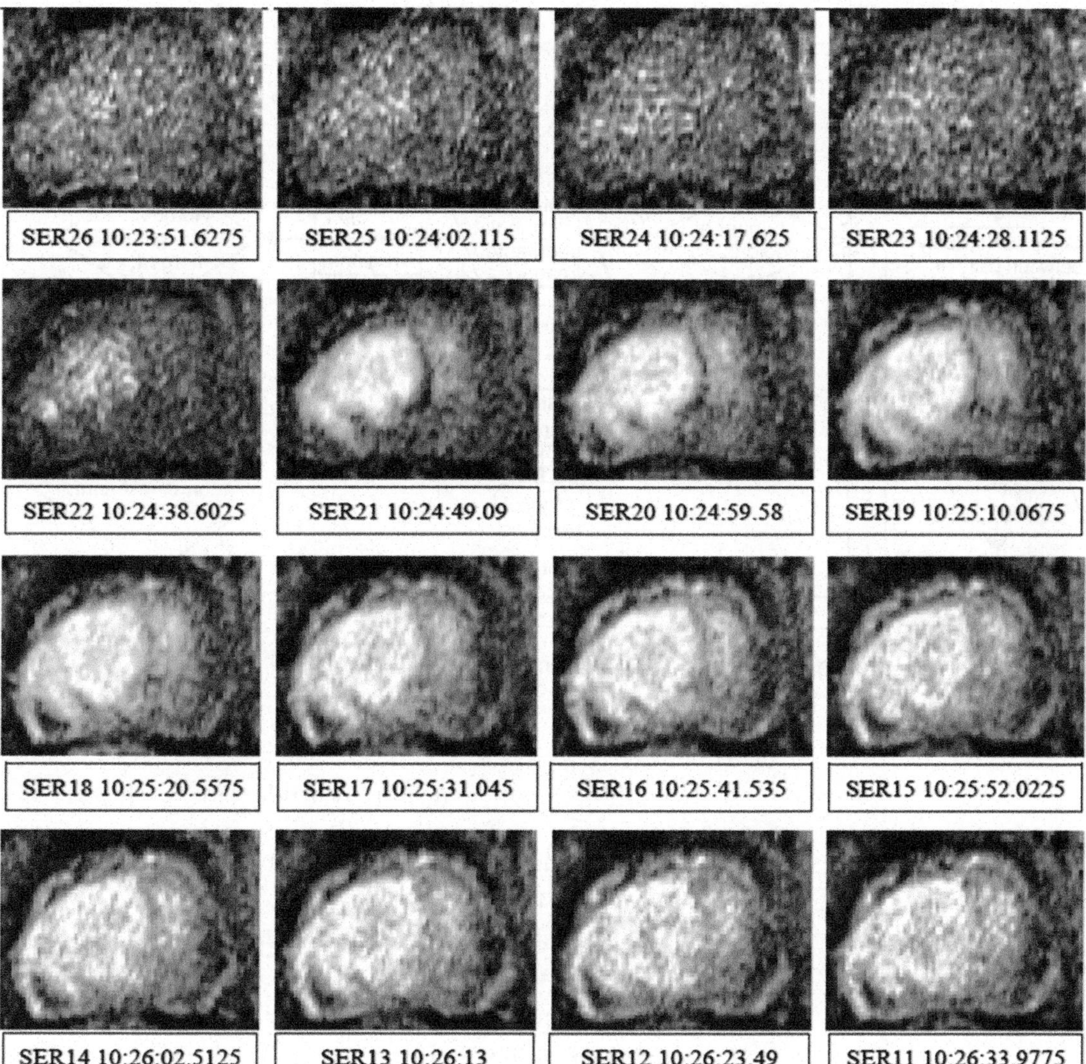

Fig. 6 Temporal sequence of the DCE MRI image of a section of the Prostate

segmented images is generated by partitioning an image into its constituent regions at hierarchical levels of allowable dissimilarity between its different regions. At any particular level in the hierarchy, the segmentation process will cluster together all the pixels and/or regions that have dissimilarity among them less than or equal to the dissimilarity allowed for that level. Refer Fig. 7 for a flowchart representation of the HCS process.

Following is a high-level description of the HCS process (Fig. 7) [5]:

1. Give each pixel in the image a region label as follows:

   If an initial segmentation of the image is available, label each pixel according to this pre-segmentation.

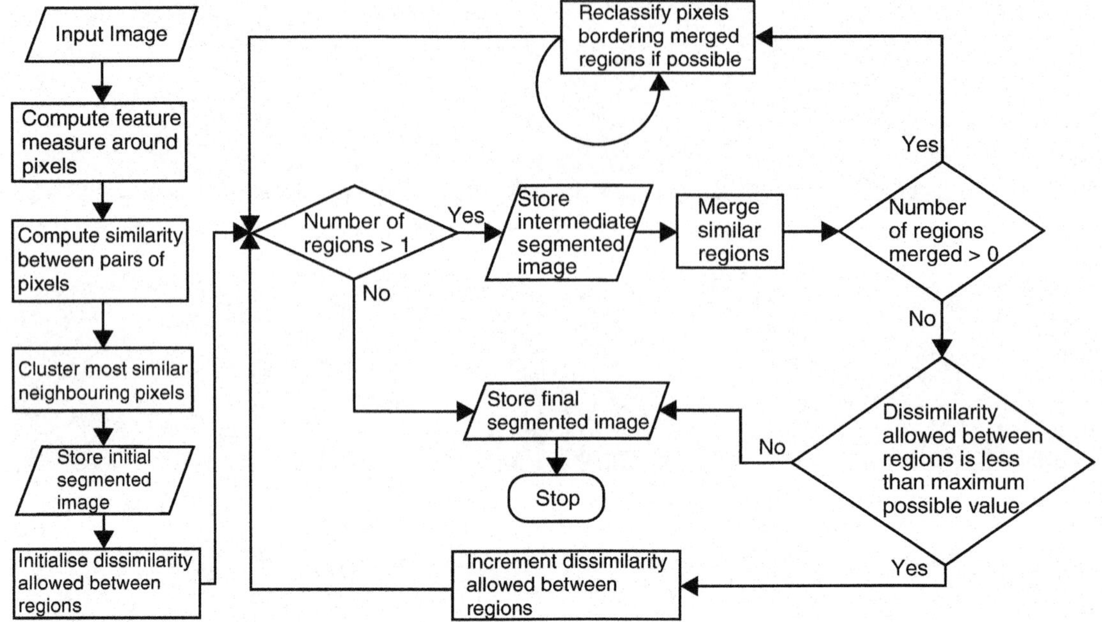

Fig. 7 Flowchart illustrating the HCS process' operational logic

If no initial segmentation is available, label each pixel as a separate region.

Set current dissimilarity allowed between regions, *dissimilarity_allowed*, equal to zero.

2. Calculate the dissimilarity value, (*dissimilarity_value*), between all pairs of regions in the image.

Set *threshold_value* equal to the smallest *dissimilarity_value.*

3. If the *threshold_value* found, in **step 2**, is less than or equal to the current *dissimilarity_allowed,*

   then merge all those regions having *dissimilarity_value*, between them, less than, or equal to the *threshold_value.*

   Otherwise go to **step 6**.

4. If the number of regions merged in **step 3** is greater than 0,

   then reclassify the pixels on the border of the merged regions with the rest of the regions until no more reclassification is possible.

   After all the possible border pixels are reclassified, among the merged regions, store the region information for this iteration as an intermediate segmentation and go to **step 2**.

   Otherwise, if the number of regions merged in **step 3** is equal to 0 then, go to **step 5**.

5. If the current number of regions in the image is less than the preset value, *check_no_regions*, go to **step** 7.
   Otherwise, go to **step 6**.
6. If the current value of *dissimilarity_allowed* is less than the maximum possible value, then increase the *dissimilarity_allowed* value by an incremental value, and go to **step 2**. Otherwise go to **step** 7.
7. Save the region information from the current iteration as the coarsest instance.

The above steps are processing intensive but ensure that the segmentation does not depend on the order in which the image regions are processed and the borders identified are more appropriate (*see* **Notes 4** and **5**).

3.2 HCS Process Aided Correlation of MRI and MALDI-MSI Image Data

The details of the steps followed in preparing the tissue sample and acquiring the corresponding MALDI-MSI data for this study is given in another publication [11] coauthored by the author. Following are the steps involved in correlating the MRI image data of the tissue with that of the MALDI-MSI data (Figs. 8 and 9).

- The original MRI image data was up-sampled (*see* **Notes 1** and **2**).
- The HCS process was applied within the region of interest (ROI) enclosing the tissue.
- The typical segmentation output was identified.
- The HCS segmentation output was correlated (visually) with the MALDI-MSI (Fig. 9).

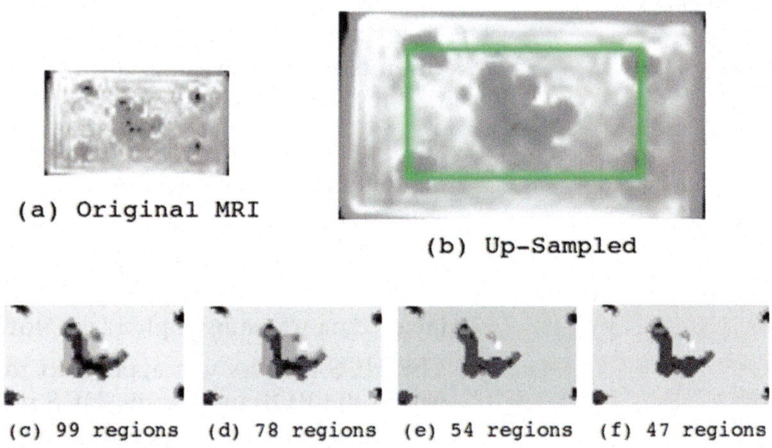

Fig. 8 Processing of the T2-weighted MRI image section (**a**) of the tissue specimen. The original image data is Up-Sampled (**b**). The HCS process relevant output highlighting the inner details of the abnormality (**c**, **d**)

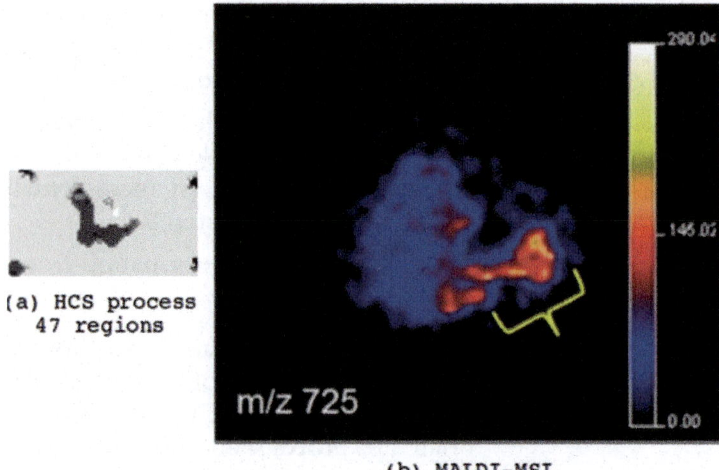

Fig. 9 Correlation of the HCS process' highlighted inner details of the abnormality with the MALDI-MS image

The primary requirement of an automated abnormalities detection system (like HCS) is the segmentation of the potential abnormal regions from noise and background. Segmentation of regions of abnormalities in images of low resolution is a challenging task. However, still, from the output in Figs. 8 and 9 one can infer that there is a potential for this approach (*see* **Note 6**).

3.3 HCS Process as a Perception Aid in Delineating Abnormalities in X-Ray Mammograms

Computer-aided detection (CAD) systems offer prompts to alert the reader to potential abnormalities. Hierarchical Clustering-based Segmentation (HCS) goes further by identifying the more appropriate edges of a lesion and heterogeneous regions within. The following method was adopted to evaluate how the HCS process outputs aid in the visualization of the details within the abnormalities in X-ray mammograms.

- Since the main aim is to aid the user to visualize the finer details of an abnormal region, to start with the approximate location and extent of the abnormality is marked by the user (Fig. 10).
- Since the original image data is of very high resolution and since the HCS process is processing intensive the original image data was subsampled (*see* **Notes 3** and **4**) (Fig. 11).
- The HCS process was applied to the subsampled image data within the ROI and for the HCS process's relevant segmentations following three types of output were created (*see* **Note 7**).

Boundary outlined dissimilar regions (Fig. 12).
Heat map of the dissimilar regions (Fig. 13).
Highlighted dissimilar regions (Fig. 14).

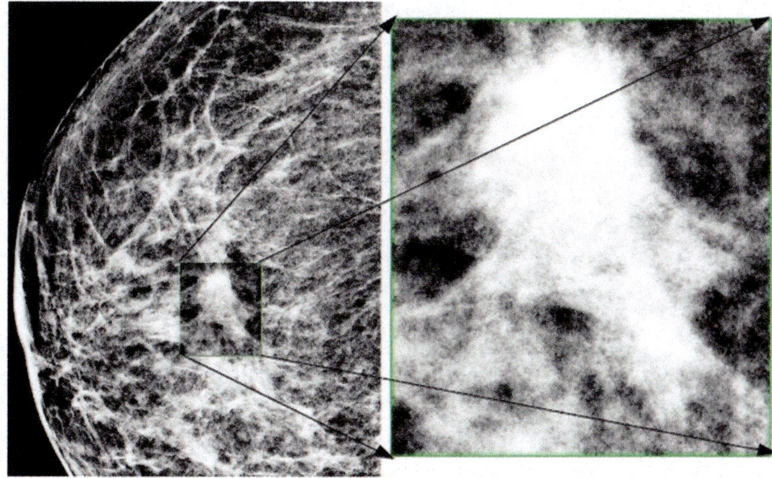

Fig. 10 Part of the original X-ray mammogram image with the part of the image having the abnormality enlarged

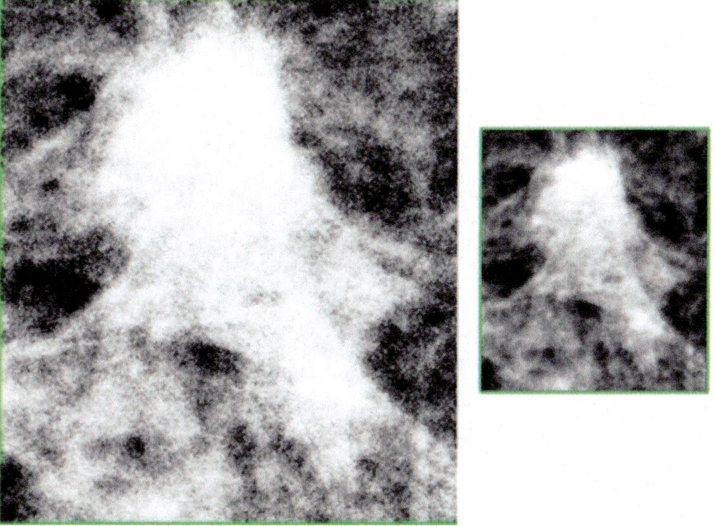

Fig. 11 Original size of the part of the image containing the abnormality and a reduced size of it after applying bilinear subsampling to the original data

We set out to determine whether the HCS process's output would be useful and offer a potential computer-based decision aid in that it can aid in the identifying of the appropriate edges of a lesion and heterogeneous regions within that lesion area.

An initial pilot study was conducted as an online study. [https://shusls.qualtrics.com/SE/?SID=SV_20oJptApsrst2xT].

The study participants were asked a few standard demographic questions and a couple of questions regarding their current use of computer-aided detection when interpreting mammograms. In the

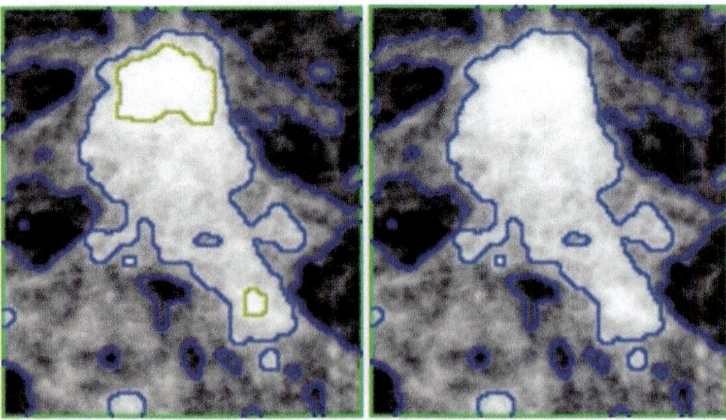

Fig. 12 HCS process' boundaries outlining the, hard to visualize, finer inner details of the abnormality

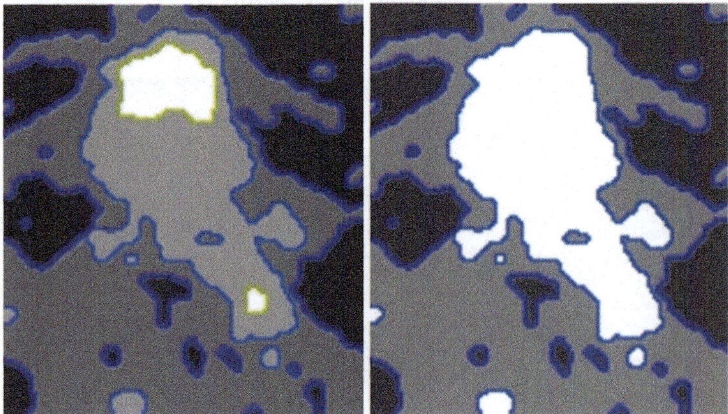

Fig. 13 Heat map image of the different dissimilar regions, within the abnormality, found by the HCS process

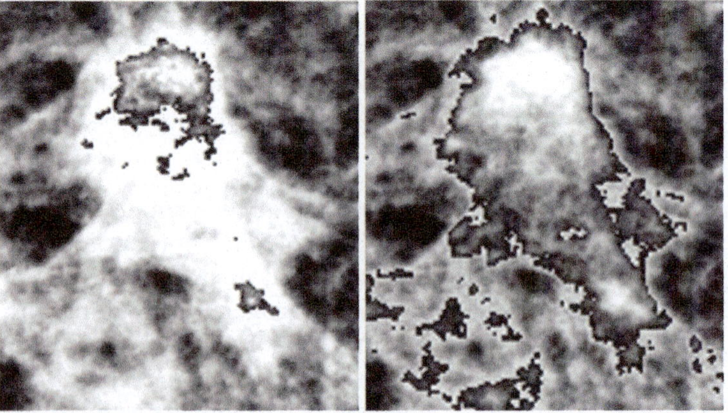

Fig. 14 Highlighted parts of the abnormality which are found to be dissimilar from the surrounding region

main task of the study the participants were presented with six mammograms, with a region of interest containing a suspected lesion already marked [12].

In condition 1, they were asked to examine this region and to indicate the extent of the lesion by using the mouse to place markers on the outer edges of the lesions. Participants were then asked whether or not they thought the lesion was multifocal and whether they thought it might be malignant.

In addition to the images shown on the screen, the original DICOM images were made available for them to download if they wish. After completing this they were then asked to repeat the task, but this time taking into consideration the additional information provided by the HCS process's output aids (condition 2) [12].

In the pilot study, the absolute differences in lesion measurements and the inter-subject reliability were compared between the two conditions. The intraclass correlation coefficient was used as an estimate of reliability and was similar between the two conditions (0.59, 0.57) [12].

When the images were divided into two groups, lesions with distinct borders and lesions with diffuse borders, there was a significant change in the absolute differences in the lesion measurements when the lesion borders were indistinct, but not in the images where the lesion borders were clearly defined.

Posthoc tests showed a significant difference in the lesion measurements when using the HCS output in cases when the lesion boundaries were indistinct ($t(2) = -7.42$, $p = 0.018$) than when the lesion boundaries were clearly defined ($t(2) = -1.54$, $p = 0.263$) [12].

HCS processing confirms the heterogeneous nature of seemingly homogeneous tissue. The adjustment of the lesion extent and increased confidence in making diagnostic interpretations by the mammogram reader in this study suggests that this information would be of practical use, particularly in cases where the lesion boundary information is ambiguous or indistinct (Fig. 15). This facilitates the accurate targeting of biopsy to the core of suspected

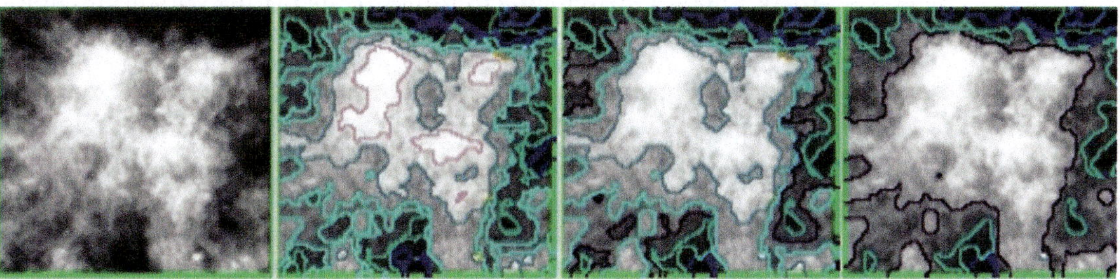

Fig. 15 Abnormality with regions having ill-defined boundaries and the HCS process' outlining of their boundaries

abnormality. Another key application is in the monitoring of tissue during and after treatment to assess the effect of drug and/or radiotherapy [12].

3.4 HCS Process Aided Interpretation of Multi-Parametric MRI (mpMRI) of Prostate

Studies have demonstrated that information from the T2-weighted MR images has a diagnostic performance such that it complements a DCE T1-weighted-based computer-aided diagnostic (CADx) system in discriminating malignant lesions from normal and benign regions [13].

Prostate Imaging and Reporting and Data System Version 2 (PI-RADS v2) recommends DCE should be included in all prostate mpMRI exams so as not to miss some small significant cancers. The DCE data should always be closely inspected for focal early enhancement. If found, then the corresponding T2 W and DWI images should be carefully interrogated for a corresponding abnormality.

From the above discussion, it can be inferred that mpMRI plays a key role in the detection of Prostate cancer. In this section, we will discuss how the HCS process aids in the interpretation of mpMRI of the Prostate.

Following are the steps that are involved in the HCS process-aided interpretation of mpMRI of the prostate:

- Time Intensity Curves (TIC)-based evaluation of DCE MRI Images.
- HCS process analysis of the corresponding T2-weighted MR image section.
- Correlating the HCS highlighted regions in the T2-weighted MRI image section, with the DCE-MRI TIC-based classification.

Each of the above steps will be discussed in detail below.

3.4.1 Time Intensity Curves (TIC)-Based Evaluation of DCE MRI Images

In prostate cancer, the leaky characteristics of the tumor angiogenesis are demonstrated in DCE-MRI by the early rapid high enhancement just after the administration of contrast medium followed immediately by a relatively rapid decline. In comparison there will be a slower and continuously increasing enhancement for normal tissues [14]. The visual analysis of DCE-MRI data makes use of the above phenomena. However, the visual assessment is inherently subjective.

The above characteristics can also be demonstrated by the quantitative measurement of signal enhancement in DCE- MRI with time. The characteristic shapes of the Time intensity curves (TIC) (Fig. 16) may be used for supporting diagnosis (*see* **Note 8**).

To categorize the TIC the shape of the TIC is analyzed as follows:

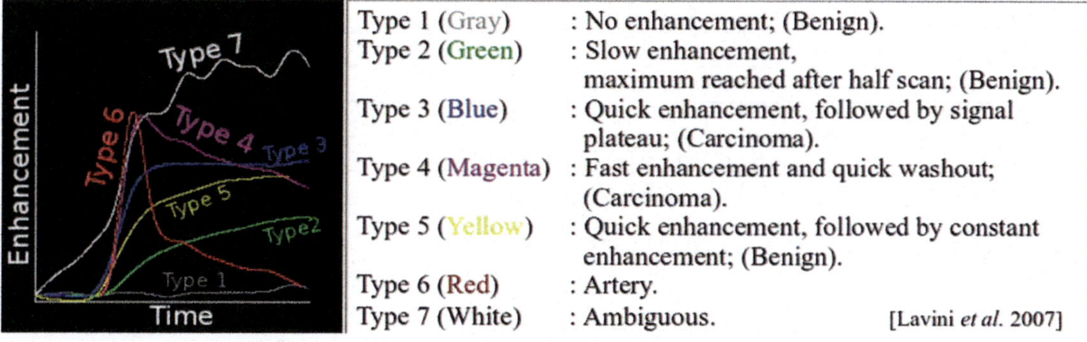

Fig. 16 Different shapes of the Time Intensity Curve (TIC) and their significance [15]

- The signal baseline (SB) was arbitrarily calculated as the average signal intensity of initial three time points before the positive slope occurs (tp = 0).
- The Tail of the TIC was assumed as the last three quarters of the time points after (tp = 0).
- The intercept (α) of the line fitted to the tail, with the axis crossing the time axis at the time tp = 0 and the tangent (β) of this line at the last time point was found.

To analyze the shape of the TIC the following features were used [15]

(a) ME: (MSD/SB), where MSD (Maximum signal difference) is the difference between the signal intensity at its maximum S(max) and SB.

(b) TTP: Time difference (in seconds) between the moment where the ME occurs and at (tp = 0). For increase-only TICs, the TTP is the last time point in the scan.

(c) MSI: Largest positive signal difference between two successive scans.

(d) RelFS: β/MSD. To describe the behavior of the curves in the last part of the scan: whether it is flat (RelFS =0), declining (RelFS <0), or increasing (RelFS >0).

For the DCE-MRI of the prostate the different TIC types (Fig. 16) are classified by a decision tree based on the above features and their threshold values listed in Fig. 17.

The HCS process-based TIC shape analysis starts with the HCS process applied to the user-selected section within a ROI (Fig. 18). The HCS process output provides the heat map images based on the normalized average pixel value of the various dissimilar regions and the regions' boundaries (Fig. 19). TICs of the

TIC Type	ME	TTP	MSI	Re1FS
1	< ME threshold			
2	> ME threshold	> 1/2 tp(maximum)	< MSD/2.0	> 0.25
3	> ME threshold		> MSD/2.0	–0.25 < Re1FS < 0.25 (Flat Tail)
4	> ME threshold	< 1/2 tp(maximum)	> MSD/4.0	Re1FS < –0.25 (Declining Tail)
5	> ME threshold	> 1/2 tp(maximum)	> MSD/2.0	Re1FS > 0.25 (Positive slope Tail)

Fig. 17 The values of the TIC curve parameters to differentiate the TIC and categorize them [15]

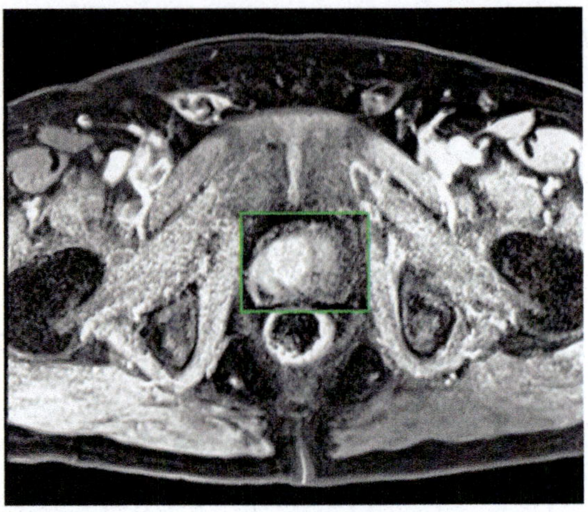

Fig. 18 DCE-MRI image section of the prostate with the region of interest (ROI) enclosing the Prostate marked. The HCS process will be applied within the ROI

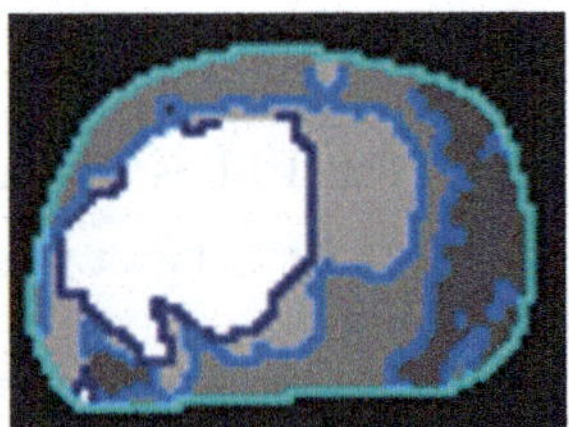

Fig. 19 Heat map image to show the different regions identified by the HCS process. The regions are shaded (*bright to dark*) based on the average pixel value of the region. The boundaries are marked by *random unique color*

contrast wash-in, wash-out process are then plotted for suspicious regions confirmed by the user (Figs. 20, 21, and 22). All the regions in the image are classified and colored based on the type of TIC for that region (Fig. 23).

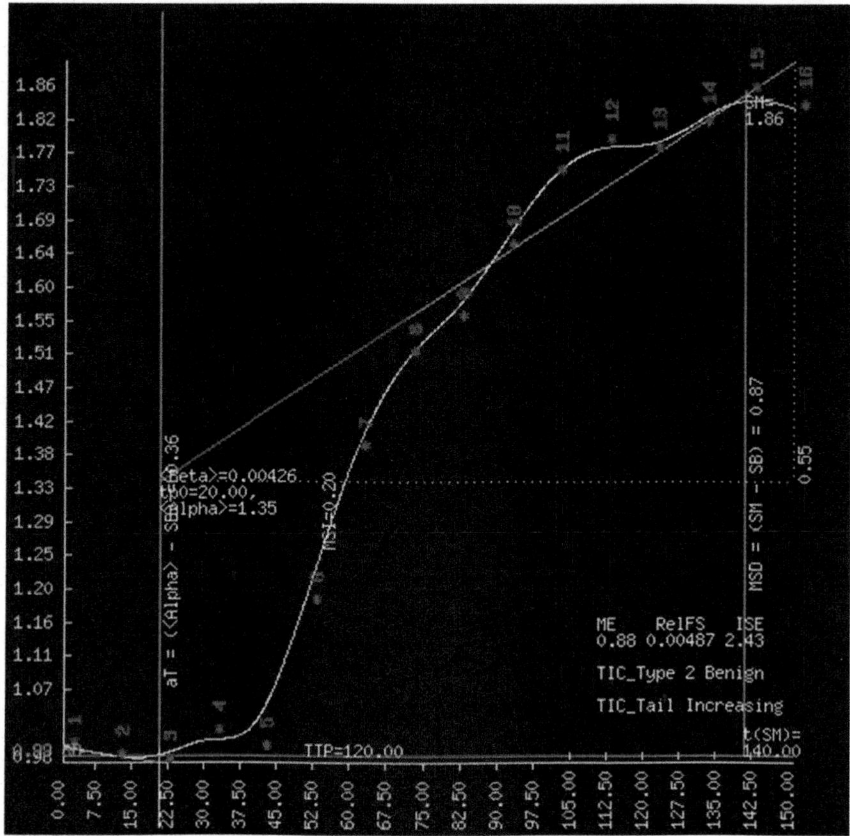

Fig. 20 Parametric illustration of the TIC of the part of the organ which is healthy. The program has automatically categorized the TIC, based on the curve's parameters, as Type 2 (Benign)

3.4.2 HCS Process Analysis of the Corresponding T2-Weighted MR Image Section

The corresponding T2-weighted MRI image section is identified. This can be done making use of the DICOM viewer and viewing the DCE-MRI image sections and the T2-weighted image sections side by side (Fig. 24). The HCS process is applied to the identified T2-weighted image section (Fig. 25) to process only the Prostate part of the section (Fig. 26).

From the HCS process output the relevant segmentation's region image (Fig. 27) and the boundary image (Fig. 28) are identified.

3.4.3 Correlating the HCS Highlighted Regions in the T2-Weighted MRI Image Section, with the DCE-MRI TIC-Based Classification

The HCS process's segmentation output of the T2-weighted image has outlined an area of signal loss (dark) (Fig. 28). Tumor in T2-weighted will appear as signal loss. But false positives occur in hemorrhage, calcification, inflammation, and fibrosis [post-inflammatory, postoperative, posthormonal (ADT), post-radiation, or following thermal ablation treatment].

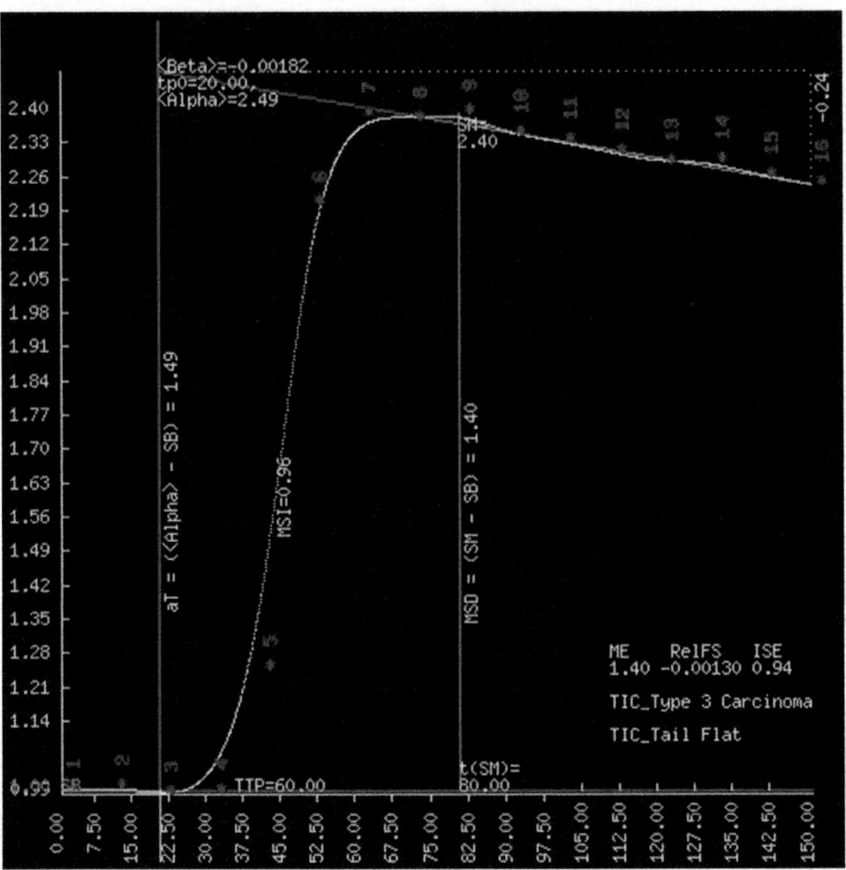

Fig. 21 Parametric illustration of the TIC of the part of the organ which has got a malignancy. The program has automatically categorized the TIC, based on the curve's parameters, as Type 3 (carcinoma)

So to confirm whether the signal loss in the T2-weighted image is due to tumor, the DCE-MRI TIC-based classification is made use of. In this case, the corresponding region, where there is signal loss in the T2-weighted image, is classified as Type-3 carcinoma (Blue) (Fig. 23).

Thus by correlating HCS boundary outlined T2-weighted image with the DCE-MRI TIC-based classified regions (Fig. 29), it can be confirmed that the signal loss in T2-weighted image is indeed due to tumor.

To ease the viewing of the different HCS segmentation output interactively and view both the T2-weighted image and the DCE-MRI image side by a GUI like the one shown in Fig. 30 can be used (Fig. 30).

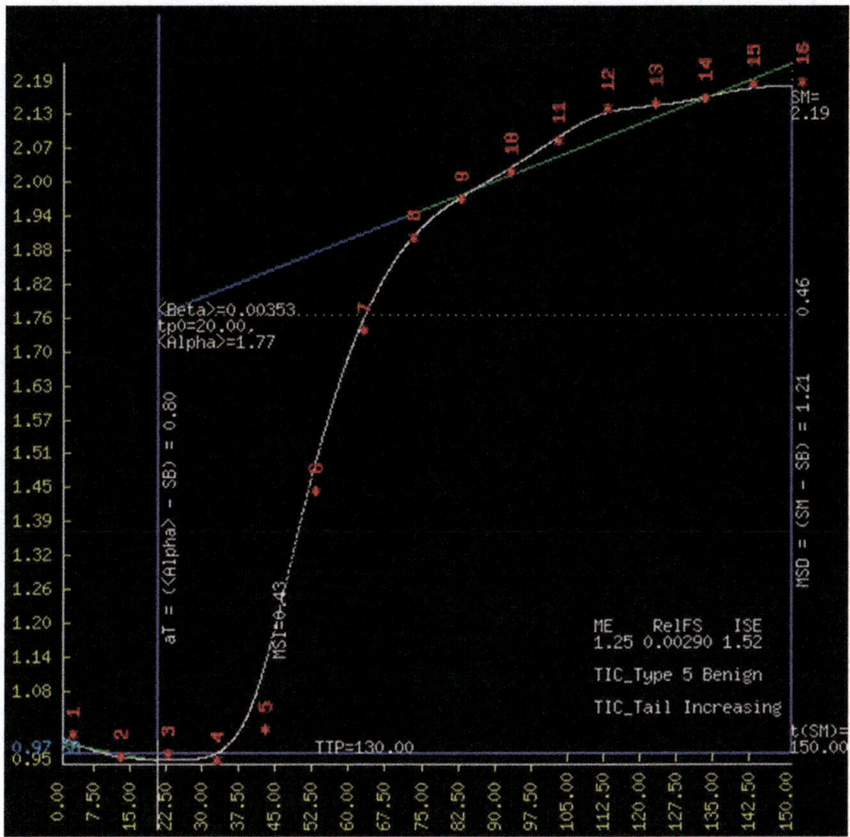

Fig. 22 Parametric illustration of the TIC of the part of the organ which is healthy. The program has automatically categorized the TIC, based on the curve's parameters, as Type 5 (Benign)

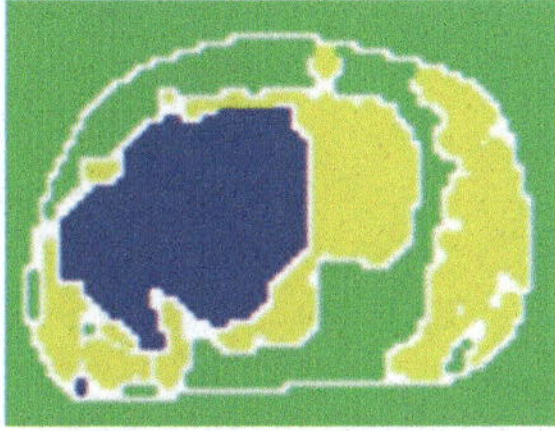

Fig. 23 The HCS process identified regions categorized as tumor (*Blue*) or healthy (*Yellow*, *Green*) based on the enhancement pattern (TIC shape) of the pixels within the region. The color coding is based on the six different types of enhancement patterns (Fig. 16)

4 Notes

1. The images were acquired by an MR equipment (with permanent magnet) having a magnetic field strength of 0.25 T and the pixel resolution was 500-μm (0.5-mm). In comparison diagnostic MRI equipment (equipped with super-conductive magnets) normally has a magnetic field strength ranging from 1.5 to 3 T.

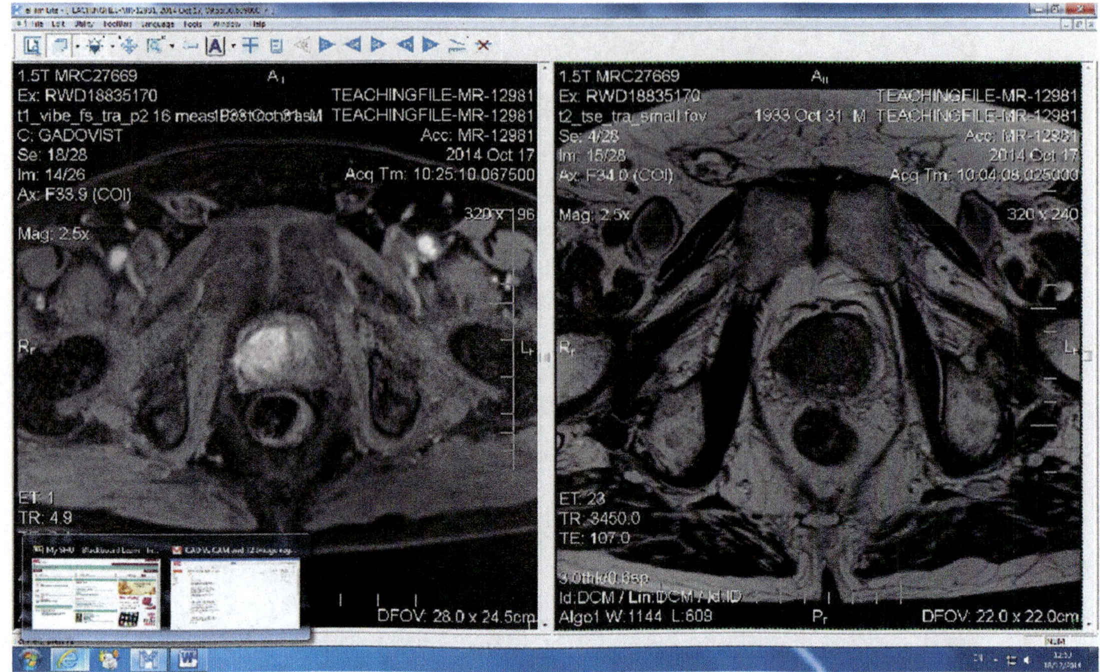

Fig. 24 Vendor provided DICOM viewer to display different image slices side by side. Using this facility it is easy to identify matching T2-weighted and the corresponding DCE MRI image of the same section

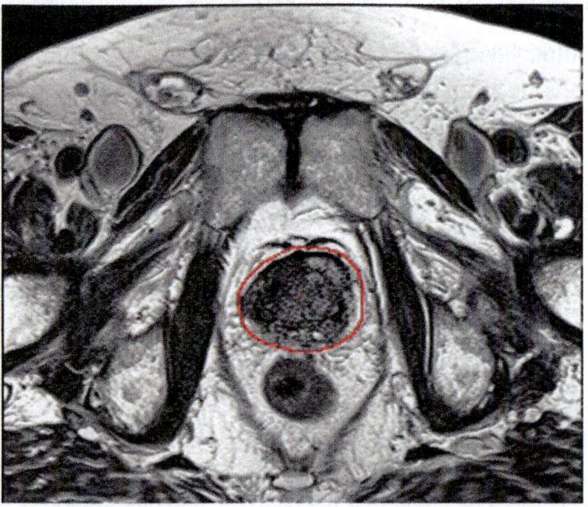

Fig. 25 T2-weighted MRI image section of the prostate with the prostate part of the image outlined. The HCS process will be applied within the boundary

2. Normally, diagnostic images are used by the radiologists to isolate boundaries around a tumor but in this case an attempt was made to visualize the details within a tumor. Also because of the lower resolution of the acquired MRI image data there were not enough pixels to resolve the finer details within the

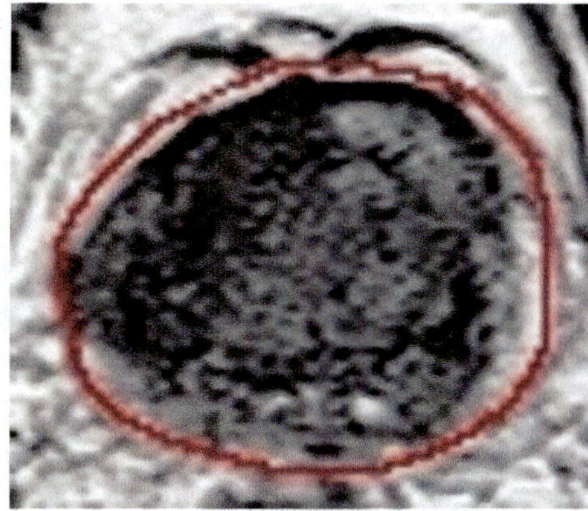

Fig. 26 The part of the T2-weighted MRI image which contains the prostate organ. The HCS process will be applied to this part of the image

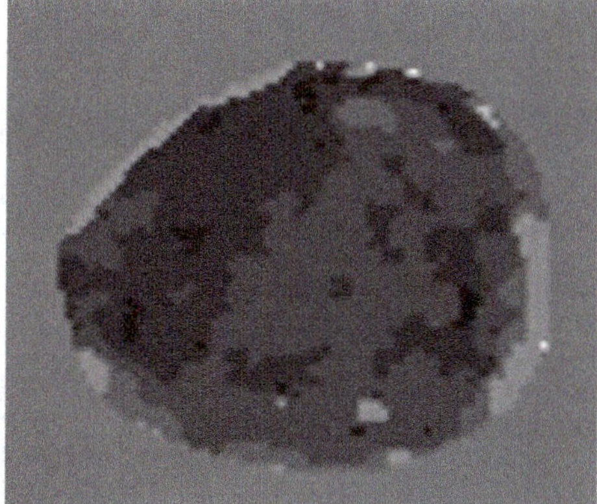

Fig. 27 Heat map image showing the different regions identified by the HCS process in processing T2-weighted MRI image. The different regions are shaded (*dark* to *bright*) based on the average pixel value of the region

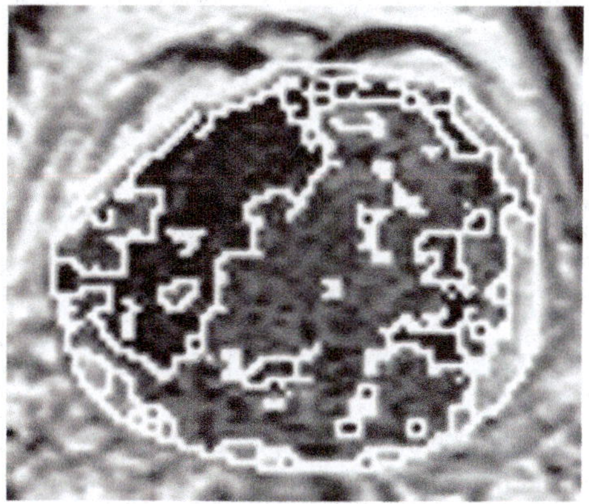

Fig. 28 The boundaries of the different regions identified by the HCS process in processing T2-weighted MRI image

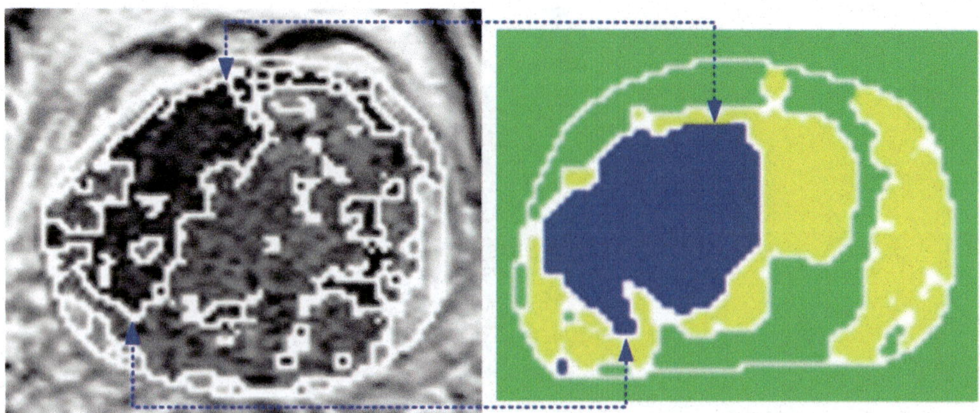

Fig. 29 Correlating HCS boundary outlined T2-weighted image (*left*) with the DCE-MRI TIC-based classified regions (*right*). The loss of signal in the T2-weighted image (*left*) matches with the DCE-MRI T2 region classified as carcinoma (*Blue region*), based on the six different types of enhancement patterns (Fig. 16)

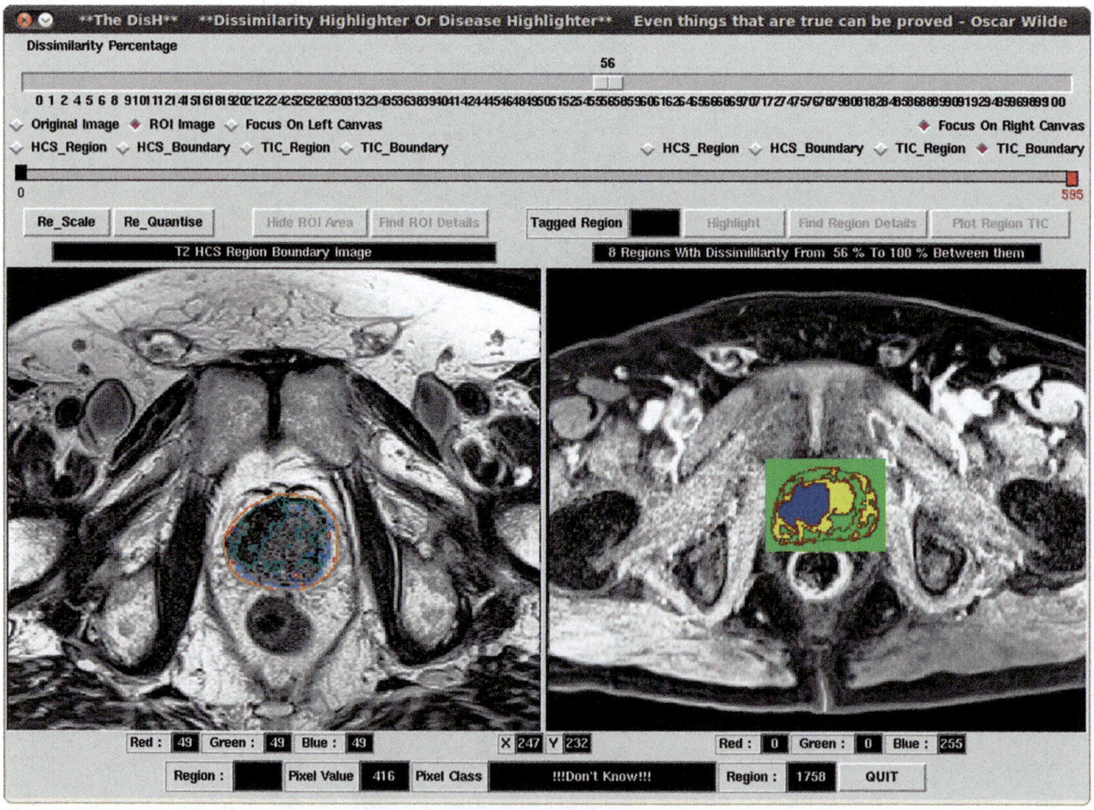

Fig. 30 Graphical User Interface (GUI) implemented as part of the study to view T2-weighted along with the DCE-MRI image slices to correlate the different regions identified by the HCS process

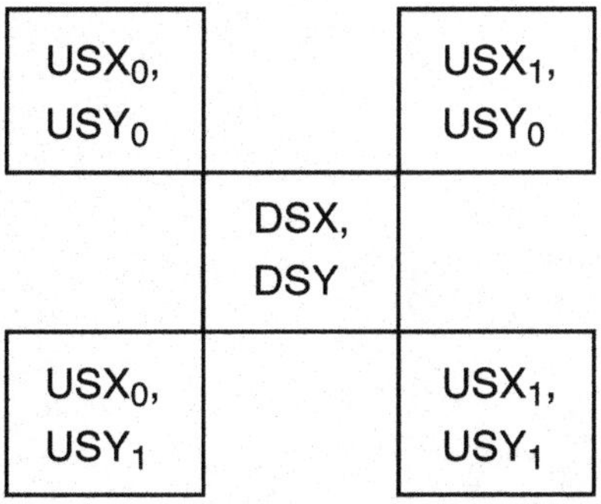

USX_0, USY_0		USX_1, USY_0
	DSX, DSY	
USX_0, USY_1		USX_1, USY_1

Fig. 31 The locations of the pixels in an image whose values will be used either to Up or Down sample the original image (For details see the text.)

tissue sample. To address this issue of sparsity of data in the original MR image, the original MR image data was Up-Sampled to twice its original size using Bi-Non-linear process (along both x and y directions). The Bi-Non-linear process is detailed below with an example.

Each of the pixels in the original image at the location (DSX, DSY) was replaced by four pixels in the Up-Sampled image at the locations (USX_0, USX_0), (USX_1, USY_0), (USX_0, USY_1), and (USX_1, USY_1) (Fig. 31).

The pixel values in the Up-Sampled image for the above four locations are estimated making use of the pixel values in the original image as follows:

$$[USX_0, USX_0] = \{9 \times [DSX, DSY] + 3 \times [DSX, (DSY-2)] + 3 \times [(DSX-2), DSY] + [(DSX-2), (DSY-2)]\} \div 16$$

$$[USX_1, USX_0] = \{9 \times [DSX, DSY] + 3 \times [DSX, (DSY-2)] + 3 \times [(DSX+2), DSY] + [(DSX+2), (DSY-2)]\} \div 16$$

$$[USX_0, USX_1] = \{9 \times [DSX, DSY] + 3 \times [DSX, (DSY+2)] + 3 \times [(DSX-2), DSY] + [(DSX-2), (DSY+2)]\} \div 16$$

$$[USX_1, USX_1] = \{9 \times [DSX, DSY] + 3 \times [DSX, (DSY+2)] + 3 \times [(DSX+2), DSY] + [(DSX+2), (DSY+2)]\} \div 16$$

Figure 32 illustrates the performance of the implemented Bi-Non-Linear Up-Sampling process where a 256 × 256 original image was Up-Sampled to a 512 × 512 image.

3. The original FFDM X-ray mammogram data was of high resolution (70 μm pixel spacing). The HCS process is a processing intensive process (*see* **Note 4**). Hence to process the images in a reasonable time it was decided to down-sample the original image data to half its original resolution (along both x and y directions). For Down-Sampling the original image data Bi-Non-linear process as detailed below was adopted.

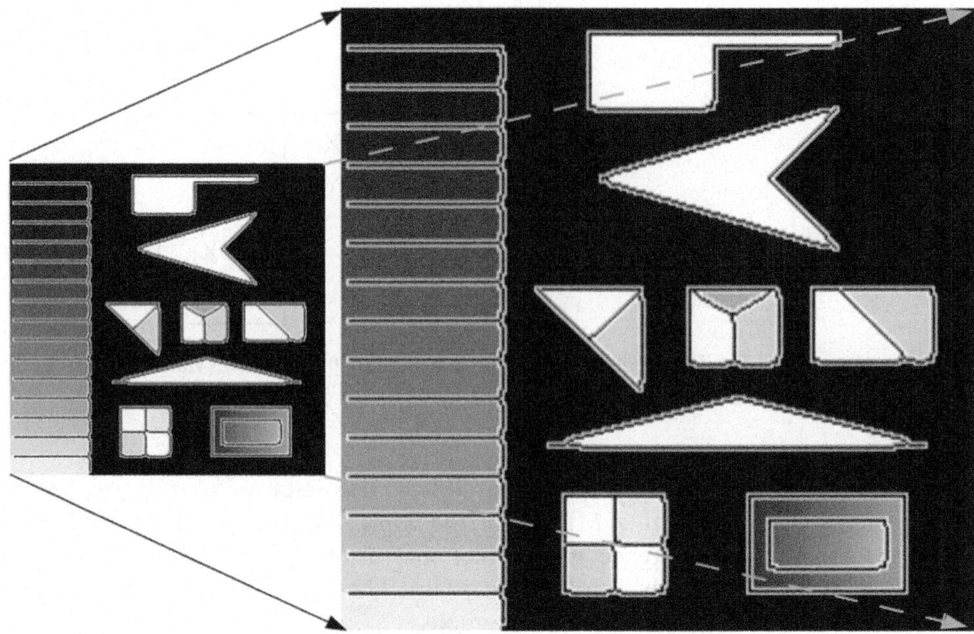

Fig. 32 Image illustrating the performance of the bilinear up-sampling algorithm

The four pixels in the original image at the locations (USX_0, USX_0), (USX_1, USY_0), (USX_0, USY_1), and (USX_1, USY_1) were replaced with a pixel at the location (DSX, DSY) in the down-sampled image (Fig. 31).

The pixel value in the Down-Sampled image is estimated making use of the pixel values in the original image as follows:

$$\begin{aligned}
&[DSX, DSY] = \{[(DSX-1),(DSY-1)] + 3\times[DSX,(DSY-1)] + 3\times[(DSX+1),(DSY-1)] + \\
&[(DSX+2),(DSY-1)] + 3\times[(DSX-1),DSY] + 9\times[DSX,DSY] + \\
&9\times[(DSX+1),DSY] + 3\times[(DSX+2),DSY] + 3\times[(DSX-1),(DSY+1)] + \\
&9\times[DSX,(DSY+1)] + 9\times[(DSX+1),(DSY+1)] + \\
&3\times[(DSX+2),(DSY+1)] + [(DSX-1),(DSY+2)] + \\
&3\times[DSX,(DSY+2)] + 3\times[(DSX+1),(DSY+2)] + [(DSX+2),(DSY+2)]\} \div 64
\end{aligned}$$

Figure 33 illustrates the performance of the implemented Bi-Non-Linear Down-Sampling process where a 512 × 512 original image was Down-Sampled to a 256 × 256 size image.

4. Other similar agglomerative hierarchical clustering or bottom-up methods suffer from the distorting phenomena, in which the cluster structures depend on order in which the regions are considered for merging [10]. But HCS process ensures that the segmentation of image into its constituent regions is always the same irrespective of the order in which image regions are processed [5]. This is achieved by the brute force approach, followed by the HCS process, where only those

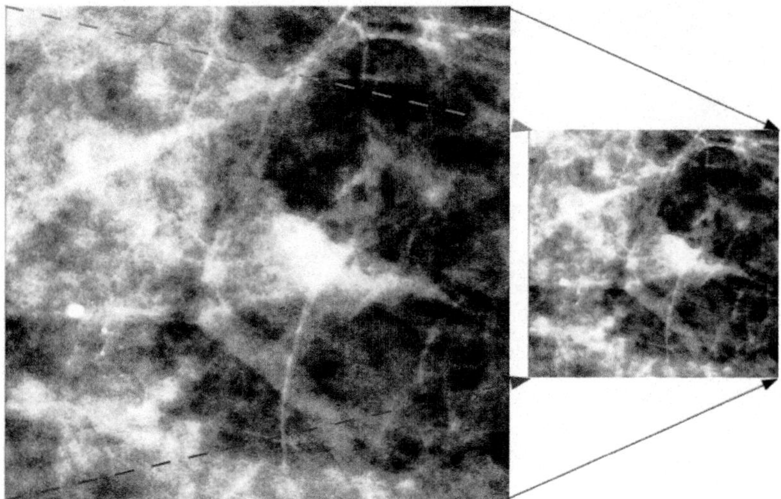

Fig. 33 Image illustrating the performance of the bilinear down-sampling algorithm

regions with the smallest overall dissimilarity are merged in each step, which is the only solution to overcome the distorting phenomena and achieve the same segmentation output consistently [16]. To accelerate the process of comparison of different regions in an image, the operation is done concurrently.

Border pixel reclassification is another unique feature of the HCS process (Fig. 7). Border pixel reclassification is considered only for those pixels on the boundary of the clusters which had been merged with other clusters. These boundary pixels are removed one at a time from their original clusters. The pixel removed is considered as a region of its own and the similarity between the one pixel region and the regions bordering it (which include the original cluster to which it belonged) is found and the single pixel region merged with the most similar bordering region. Border pixel reclassification aides in overriding local inhomogeneity while clustering similar pixels/regions. The positive effect of border pixel re-classification can be visualized in Fig. 34. It can be seen from the border outlined images the HCS process with border pixel reclassification (top row) achieves far better results in delineating the different regions of an ill-defined abnormality presented in a X-ray mammogram (middle row) when compared to the segmentation without border-pixel-re-classification (bottom row) (Fig. 34).

Because of the brute-force approach adopted by the HCS process, to ensure consistent segmentation and because of the border-pixel-re-classification operation which is a sequential operation the HCS process is a processing intensive process.

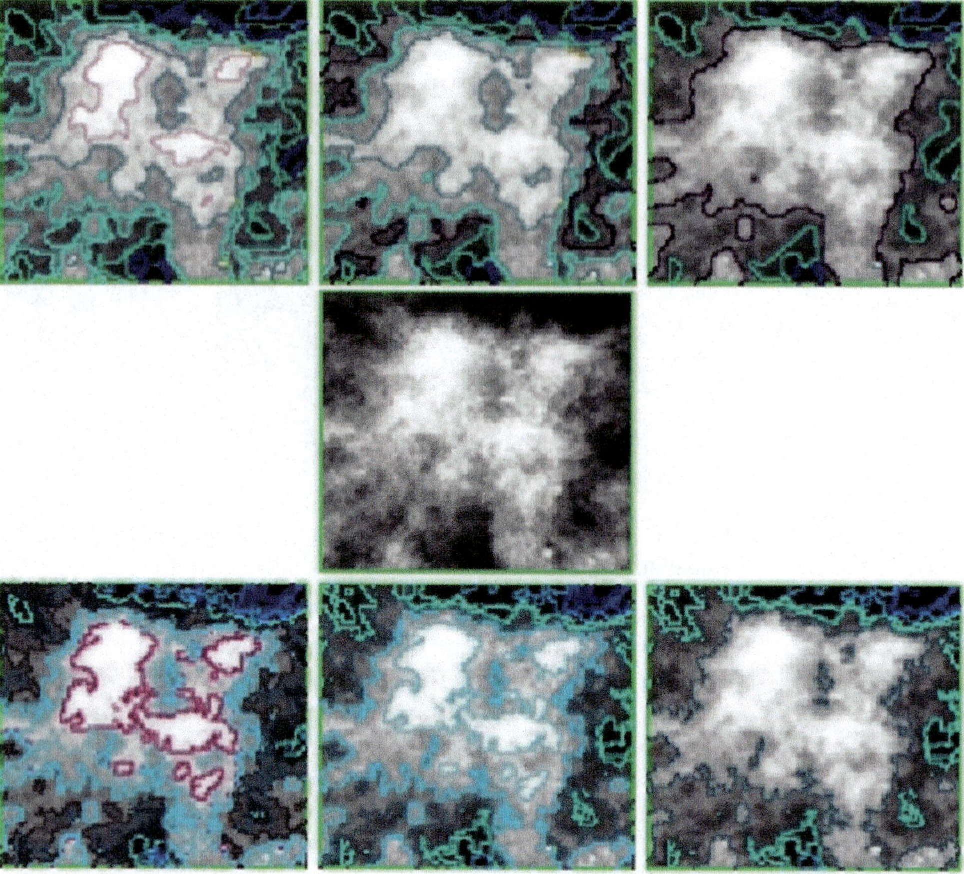

Fig. 34 An abnormality having ill-defined boundaries (*middle row*) processed by the HCS process with border pixel reclassification operation (*top row*) and without border pixel reclassification (*bottom row*). This is to illustrate how border pixel reclassification operation ensures more appropriate delineation of the ill-defined inner details of the abnormality

5. The feature measure used by the HCS process, to estimate the similarity between locations or regions within an image, is the actual distribution of the pixel values within a neighborhood surrounding the locations or within the regions. This technique may be considered to work in a way similar to the human visual system where features for texture (region) segmentation are not consciously computed [17].
6. The major limitation of the current study is that the acquired MRI image data was of low spatial resolution of only 500 μm (*see* **Notes 1** and **2**), while the MALDI-MSI image was acquired using raster/spot to spot imaging mode at 100 μm spatial resolution. However, still, HCS process segmented abnormal regions correlated with the MALDI-MSI. Hence, one can only expect improvements in the current results if

image data from higher field strength device, capable of higher resolutions like 100 μm, is used.

7. Even though the original FFDM data was down-sampled before processing, the finer details of the image was still preserved. This is because to start with the original image data was of high resolution (70 μm) and subsampling resulted in an image of 140 μm pixels spacing that is still of high resolution.

   Since the HCS process is robust enough to handle image data of very low resolution (250 μm), the processing of the down-sampled image data (140 μm) still facilitated visualization of finer details within the abnormality (Fig. 12).

8. To interpret the DCE-MRI data, by making use of TIC characteristics, machine vendors provide the radiologists with the facility to select a region of interest (ROI) enclosing an area of the largest enhancement and subsequently observe how the average signal intensity of the voxels within the ROI varies with time (Fig. 35).

   The ROI, normally chosen within an area of the largest enhancement, because of tissue heterogeneity, may enclose tissues of different enhancement patterns (Green boundary Fig. 36a). Hence, the averaged TIC from the ROI may not represent the actual characteristics of the lesion.

   To overcome the approximation, intrinsic to the TIC estimated on pixels averaged within the radiologist drawn ROI, recent studies have proposed to estimate and classify the TIC in every single voxel acquired by the DCE-MRI scan sequence [18].

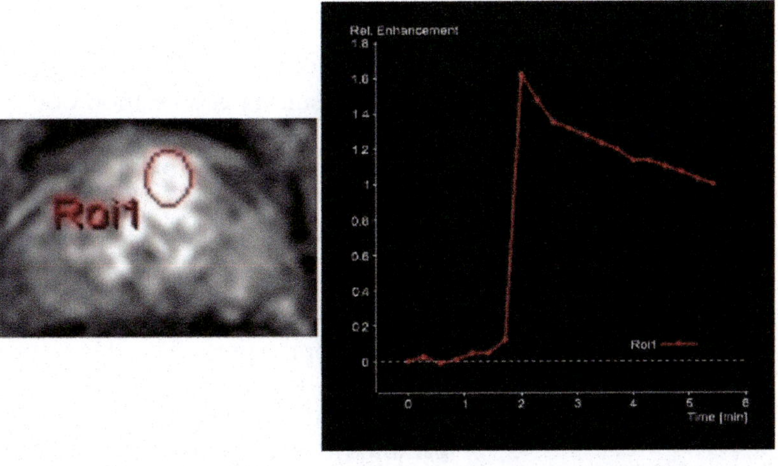

Fig. 35 Region of interest (ROI) selected by the radiologist, enclosing an area of the largest enhancement and the corresponding enhancement pattern, of the locations within the ROI, with time (TIC). (This facility is provided by the MRI machine vendor)

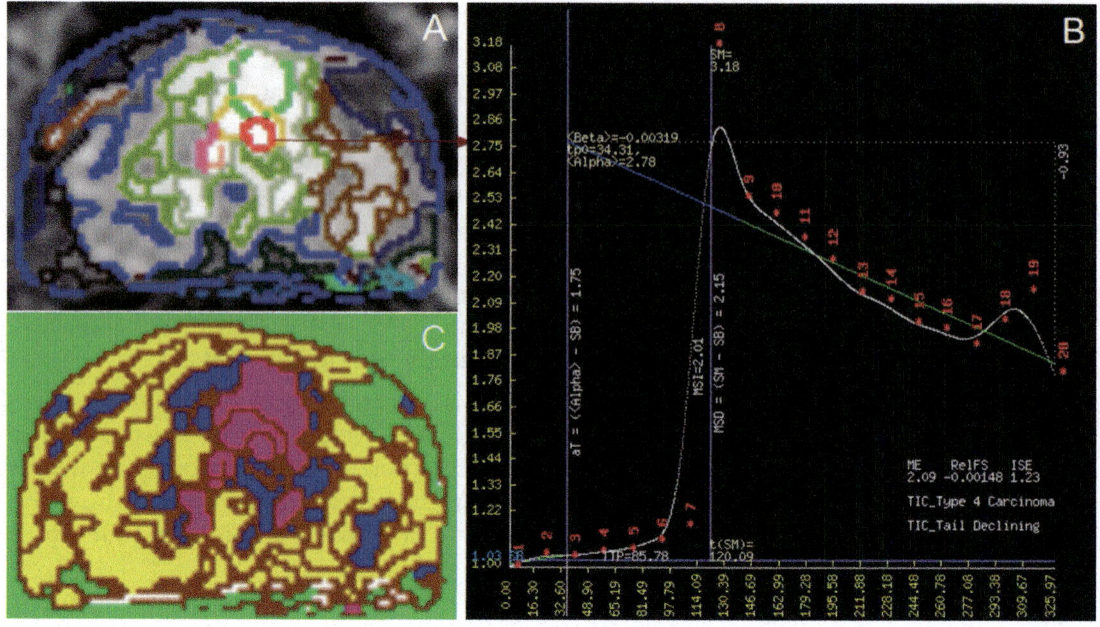

Fig. 36 ROI selected by the radiologist, enclosing tissues of different enhancement patterns (*Green boundary*) (**a**). Making use of the HCS process segmented regions user objectively chooses a more appropriate ROI (*Red boundary*) (**a**). Parametric illustration of the TIC for the user chosen ROI (**b**). Automated color-coded classification, of the HCS process' different regions, based on the TIC enhancement pattern of the regions (**c**). (The color coding is based on the six different types of enhancement patterns (Fig. 16))

The limitations of the pixel by pixel analysis of the TIC are as follows:

- Pixel-by-pixel analysis of the TIC is sensitive to pixels having different enhancement patterns [19].
- Pixel-by-pixel analysis of the TIC excludes the user from the diagnostic process.
- The resulting classification does not provide any indication regarding how and why pixels are classified as belonging to a specific type which might lead to the situation where incorrect CAD can have a detrimental effect on human decisions [20, 21].

The demonstrated HCS process-based TIC classifier method offers the benefits of both the ROI and pixel-by-pixel analysis approaches. The HCS process segmented regions enable the user to objectively choose a more appropriate ROI (Red boundary Fig. 36a) and to view a more representative parametric illustration of the TIC (Fig. 36b).

Also, the HCS process-based method presents the user with the automated color-coded classification of the different regions based on the TIC enhancement pattern of the HCS process's regions (Fig. 36c). The color coding is based on the six different types of enhancement patterns (Fig. 16).

References

1. Chandarana H, Rosenkrantz AB, Mussi TC, Kim S, Ahmad AA, Raj SD, McMenamy J, Melamed J, Babb JS, Kiefer B, Kiraly AP (2012) Histogram analysis of whole-lesion enhancement in differentiating clear cell from papillary subtype of renal cell cancer. Radiology 265:790–798
2. Fitzpatrick JM, Sonka M (eds) (2000) Handbook of medical imaging volume 2 medical image processing and analysis. SPIE-the international society for optical engineering. ISBN-10: 0819477605
3. Arbelaez P (2006) Boundary extraction in natural images using Ultrametric contour maps. Proceedings 5th IEEE workshop on perceptual Organization in Computer Vision (POCV'06), New York, June 2006 (paper presented at a conference)
4. Arbelaez P, Fowlkes C, Martin D (2007) The Berkeley segmentation dataset and benchmark. http://www.Eecs.Berkeley.Edu/research/projects/CS/vision/bsds/. Accessed 27 Jun 2016
5. Selvan AN (2007) Highlighting dissimilarity in medical images using hierarchical clustering-based segmentation (HCS). M.Phil. Dissertation, Faculty of Arts Computing Engineering and Sciences, Sheffield Hallam University, Sheffield, UK
6. Selvan AN (2011) Boundary extraction in images using hierarchical clustering-based segmentation (HCS). Paper presented at the British machine vision conference (student workshop), Dundee, UK, Sept 2011
7. Selvan AN (2012) Hierarchical clustering-based segmentation (HCS) aided diagnostic image interpretation and monitoring. Doctoral dissertation, Faculty of Arts Computing Engineering and Sciences, Sheffield Hallam University, Sheffield, UK
8. Cole LM, Selvan AN, Wright C, Reed H, Clench MR (2014) Communication of medical images to diverse audiences using multiple imaging modalities. Poster presented in British mass spectrometry society, Cheshire, UK, 1–2 April 2014
9. Ohlander R, Price K, Reddy R (1978) Picture segmentation by a recursive region splitting method. Comput Graph Image Process 8:313–333
10. Nadler M, Smith EP (1993) Pattern recognition engineering. John Wiley and Sons, New York
11. Cole LM, Selvan AN, Partridge R, Reed H, Wright C, Clench MR (2015) Communication of medical images to diverse audiences using multimodal imaging. Adv Struct Chem Imag 1:13
12. Naylor S, Spackman L and Selvan AN (2015) Evaluation of hierarchical clustering-based segmentation as a perception aid for mammogram readers. Poster presented in UK radiological congress, ACC, Liverpool, UK, 29 June–1 July 2015
13. Vos CP, Hambrock T, Barenstz JO, Huisman HJ (2010) Computer-assisted analysis of peripheral zone prostate lesions using T2-weighted and dynamic contrast enhanced T1-weighted MRI. Phys Med Biol 55(6):1719–1734
14. Engelbrecht MR, Huisman HJ, Laheij RJ et al (2003) Discrimination of prostate cancer from normal peripheral zone and central land tissue by using dynamic contrast-enhanced MR imaging. Radiology 229:248–254
15. Lavini C, de Jonge MC, van de Sande MG, Tak PP, Nederveen AJ, Maas M Pixel-by-pixel analysis of DCE MRI curve patterns and an illustration of its application to the imaging of the musculoskeletal system. Magn Reson Imaging 25(5):604–612
16. Thies C, Malik A, Keysers D, Kohnen M, Fischer B, Lehmann TM (2003) Content-based retrieval in medical image databases by hierarchical feature clustering. Proc SPIE 5032:598–608
17. Bhattacharya U, Chaudhuri BB, Parui SK (1997) An MLP-based texture segmentation method without selecting a feature set Image and Vis Comput 15 937–948
18. Lavini C et al (2007) Pixel-by-pixel analysis of DCE MRI curve patterns and an illustration of its application to the imaging of the musculoskeletal system. MRI 25(5):604–612
19. Lavini C, Buiter SM, Maas M (2013) Use of dynamic contrast enhanced time intensity curve shape analysis in MRI: theory and practice. Rep Med Imag 2013:71–82
20. Alberdi E et al (2005) Use of computer-aided detection (CAD) tools in screening mammography: a multidisciplinary investigation. Br J Radiol 78:31–40
21. Fenton JJ et al (2011) Effectiveness of computer-aided detection in community mammography practice. J Natl Cancer Inst 103(15):1152–1161

Chapter 11

Laser Ablation Inductively Coupled Plasma Mass Spectrometry Imaging of Plant Metabolites

Callie Seaman

Abstract

LA-ICP-MS (Laser Ablation Inductively Coupled Plasma Mass Spectrometry) is a highly sensitive isotopic and elemental analytical technique that has a wide application for solid samples such as plant material. Within this chapter we describe how to prepare leaf material, seeds, and fruit for 2D elemental distribution maps, preparation of reference materials, method condition optimization, special resolution calculation, and the limit of detection.

Key words Laser ablation, ICP-MS, Imaging, Reference material, Resolution, Method optimization, Calibration, Quantification

1 Introduction

Laser Ablation Inductively Coupled Plasma Mass Spectrometry (LA-ICP-MS) is a highly sensitive isotopic and elemental analytical technique that has a wide application for solid samples such as plant material. It offers the ease of direct analysis of the sample, with minimal sample preparation and the size of the sample needs only to be picograms to femtograms in mass size. It has a versatile application for sampling schemes including depth profiling, bulk analysis, defect analysis, and both elemental and isotopic mapping at levels down to ppb (parts per billion).

The process of LA-ICP-MS, as the name suggests, begins with a laser beam, such as a NdYAG UV laser (213 nm), being focused onto the surface of a sample to produce fine particles. These particles are then transported by a carrier gas, such as argon, to the plasma of an ICP-MS instrument, to be digested and ionized before entering a mass spectrometer detector where they are analyzed for isotopic and elemental content. To combat isobaric interference that is experienced by many isotopes, the use of a collision cell can be adopted, using helium as the collision gas.

Laura M. Cole (ed.), *Imaging Mass Spectrometry: Methods and Protocols*, Methods in Molecular Biology, vol. 1618,
DOI 10.1007/978-1-4939-7051-3_11,

Quantitative imaging is achieved by spiking matrix-matched reference material, quantifying them and then running alongside the tissue for imaging. The signal produced by the reference materials is then compared against the signal produced in the image.

Within this chapter we describe how to prepare leaf material, seeds, and fruit for 2D elemental distribution, preparation of reference materials, method condition optimisation, special resolution calculation, and the limit of detection.

2 Materials

2.1 Tissue Cryosectioning and Embedding

1. Propanol, cooled using liquid nitrogen.
2. Liquid nitrogen for snap freezing (seeds/fruit), in a suitable insulated container.
3. Embedding mediums; 1% carboxymethylcellulose (CMC) or 5% gelatine.

2.2 Chemicals and Materials for Imaging

1. Adhesive microscope slides for tissue mounting.
2. 25% w/w powdered vanillic acid to bind the standards.
3. Carbon tape to secure standard array.

3 Method

3.1 Sample Preparation

1. Plant material for sectioning such as seeds or fruit are snap frozen. Only leaves for reference material production are snap frozen, as leaf tissue loses its shape and becomes very brittle.
2. Fill a small strong plastic beaker with enough propanol to cover the tissue. Do not use glass as it has a tendency to shatter and crack when placed into liquid nitrogen.
3. Take a separate beaker, large enough to fit the small beaker of propanol in, and place the beaker of propanol within.
4. Fill the larger beaker with liquid nitrogen to cool the propanol. Make sure that the liquid nitrogen does not cover the top of the propanol filled beaker.
5. The plant material is then gently placed into the precooled propanol for approximately 2 min, until all of the tissue is frozen and the tissue appears white in color.
6. Carefully remove the tissue from the propanol, place into a clean freezer safe container and store at −80 °C.

3.2 Cryosectioning and Embedding

1. Seeds are first embedded in 1% carboxymethylcellulose (CMC) or 5% gelatine and placed into −80 °C freezer overnight (*see* **Note 1**).

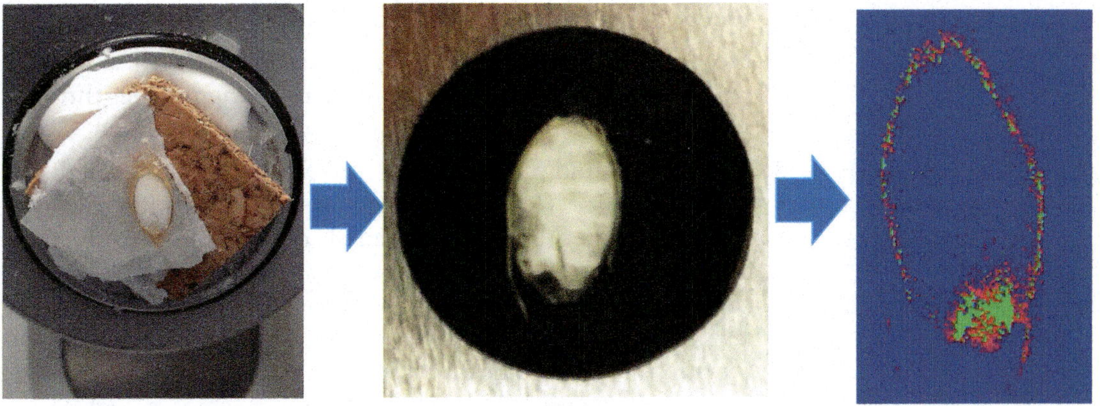

Fig. 1 Schematic diagram on (**a**) cryosection of barley seed, (**b**) mounted the section, and the (**c**) LA-ICP-MS image

2. The frozen, embedded seeds are sectioned at a temperature of −24 °C to obtain 20 μm sections, which are then mounted onto glass slides using carbon tape as the adhesive (*see* **Notes 2–4**) (Fig. 1).

3.3 Leaf Material for Imaging

1. Clean and remove any debris, first with a soft clean brush, and then with deionized water, to remove any residues, nutrients, or particles, which could provide false readings (*see* **Note** 7).
2. Fresh leaves are then placed between tissue paper and pressed between two glass slides.
3. The slides are taped tightly together with masking tape and placed into a freeze drier at −36 °C until they reach a constant mass.
4. Prepare slides by cutting double-sided carbon tape to the correct size and placing onto the slide. Ensure that the size of the leaf does not exceed the ablation area on the stage of the laser ablation system (LA).
5. Carefully un-wrap the tape, and remove the dried leaf.
6. Mount the dried leaves onto the carbon tape on the glass slides, ensuring it is as flat and even as possible. This can be achieved by placing the removed carbon tape cover over the area and pressing firmly with another slide (*see* **Note 5**).

3.4 LA Reference Material Preparation

1. Obtain fresh matrix matched plant material. It is important that it is as closely matched as possible to ensure accurate results. So, for plant leaves, use leaves from the same type of plant, and for seeds, use the same variety (*see* **Note 6**).
2. Wash the material with deionized water to remove any residual nutrients or particles and snap freeze using the method in the sample preparation section.

3. Freeze dry to a constant mass to remove any residual water.
4. Homogenize the material by grinding in an electrical grinder or with a pestle and mortar. Pass the resulting powder through a 40-μm sieve to ensure homogeneity.
5. Split the ground plant material into five groups and weigh 1 g ± 0.01 into precleaned falcon tubes. To each tube add the same volume of standard elemental solution, but with increasing concentrations. Always have a blank and only add water to this one. Ensure the powder is saturated and vortex to ensure all powder is dispersed into the solution, as dry clumps can form in the bottom of the tube.
6. Cap the tubes and place onto an oscillating table for 30 min at 200 RPM.
7. Instantly freeze the tubes containing the standards in liquid nitrogen [1], ensuring the whole sample is solid and put into the freezer at −20 °C overnight.
8. Remove from the freezer and place the standards onto a freeze drier at −52 °C and 0.120 mBar until a constant mass is achieved.
9. Add 25% w/w powdered vanillic acid to each of the dried standards. The vanillic acid is used as a binder and it also absorbs strongly at the lasing wavelength of 213 nm. It has been previously documented [2, 3] that vanillic acid not only improves the sensitivity of LA-ICP-MS analysis, but also standardizes the absorptivity of the matrix, which, in turn, improves the quality of the data.
10. To ensure homogeneity and reduce particle size even further, pour the resulting mixture into separate plastic pots with 5 mm ceramic zirconia milling material. Rotary mill for approximately 48 h at 44.6 RPM. This is to improve the homogeneity of the samples.
11. Pass the milled plant material through a sieve to remove the milling material.
12. Press the material into 13 mm pellets using a bench press under vacuum, applying 2 ton of pressure for 1–2 min. Smaller pellets can be produced, but less pressure is required.
13. Digest and quantify the elemental content of the pellets.

3.5 Method Optimization

There are many interchangeable operating parameters that require optimizing in order to produce a fast, high-resolution elemental distribution image. Table 1 summarizes the input parameters on a laser ablation system (*see* **Notes 8** and **9**).

Laser Spot Size —The size of this is selected based on the size of the tissue and the features you wish to analyze. A smaller laser spot will give a more detailed image; however, the time taken will increase.

Table 1
LA input parameters

LA parameter	Typical parameters
Laser spot size (SS)	55 μm
Laser power (LP)	28%
Repetition rate (RR)	20 Hz
Scan rate/speed (SR)	95.4 μm/s
Scan spacing	55 μm
Laser warm up time	20s
Cell Washout time	25 s
ICP-MS acquisition time (AT)	0.576 s

Laser power —It is important to test the tissue prior to analysis and establish the correct laser power. If the energy is too high, there can be too much noise in the signal, thus making the integration of the data difficult. The following steps will help to establish the optimum laser power.

1. Set the laser power at 10% and run an ablation line on the tissue and record the signal.
2. Run fresh ablation lines on the tissue, but increase the power until damage is visible but does not penetrate the material underneath.
3. Repeat this all around the tissue, increasing the laser power incrementally, recording the signal at each level.
4. Continue to increase the power until the signal no longer increases.
5. Plot the signal against the laser power, as illustrated in Fig. 2.
6. The ablation threshold is at the point where the signal starts to increase and the optimal laser power is at the roll off point of the graph.
7. It can be worth running a couple of ablations on the material the tissue is mounted onto, and establishing its ablation threshold as well. Look for elements that are also in high concentrations within this material and monitor them, so that, if the tissue is particularly thin in some areas, it can be seen that the laser power is too high.

Repartition rate —This is selected on the basis of the tissue type. Hard substances, such as seeds, will tend to require higher repartition rates than softer tissue like root hairs.

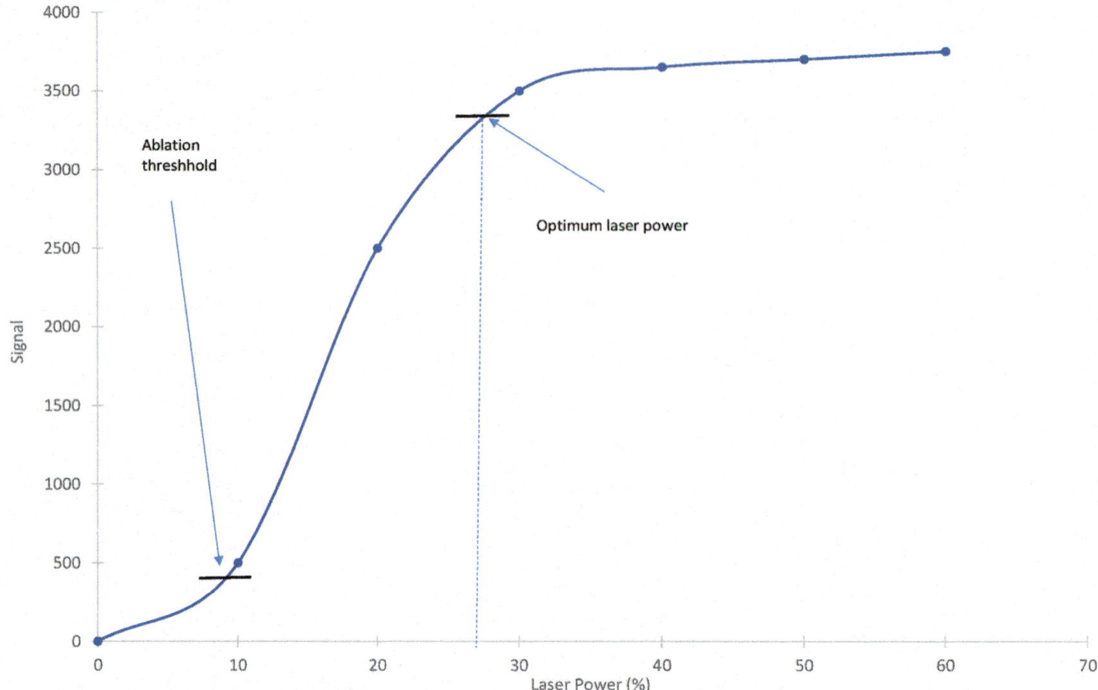

Fig. 2 Effect of laser power (%) on signal

Scan rate/speed —In order to produce high-quality images, it is very important to calculate the scan rate/speed (SR) correctly. Aim for the SR to run the same distances as the spot diameter in the acquisition time (AT) of the ICP-MS. If the scan rate/speed is too slow, the ablations will overlap, reanalyzing the same area twice. If the scan rate/speed is too fast, there are sections of the tissue which are not ablated and, therefore, make the image incomplete. Figure 3 demonstrates this. To calculate the scan rate/speed the following equation is used:

$$\text{Scan rate} = \text{Spot size} \ / \ ICP \ \text{acquisition time}$$

3.5.1 Quantification and Calibration Curve

In order to quantify an image, reference material on known elemental content must be used. Software such as IOLITE 3.32 can process this data to produce a quantitative image. Alternatively, an Excel spread sheet can be used to produce a calibration curve, with elemental concentration versus instrument signal.

1. Before analysis of the tissue, run a 60 s ablation on the reference materials, with a 30 s gas blank at the start and end.
2. As the internal standard, use ^{13}C to correct for instrument fluctuations, variation in sample thickness, and water content [4–6]. Other elements have been used in other research

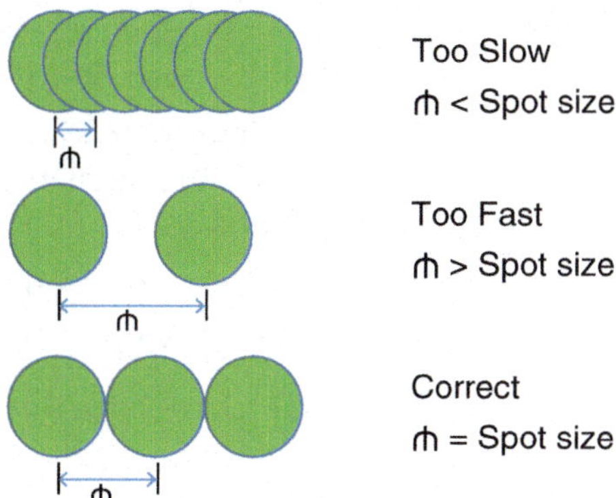

Fig. 3 The effects of scan rate/speed on data acquisition

including Sc [7] and ^{12}C [8], to compensate for the difference in the amount of material ablated [4], and results normalized against the signal.

3. After the image tissue has run, repeat the 60 s ablation on the reference materials, with a 30 s gas blank at the start and end. It is important that this line is run on a different area of the reference material.

3.5.2 Distribution Maps Image Production

Integration can be performed using data processing software such as IOLITE 3.32 or a more basic way is to use Microsoft Excel, which is very time consuming and not as accurate.

1. Prior to sample analysis, the LA-ICP-MS is optimized for *X*-*Y* torch position, lens voltage, and nebulizer gas flow by ablating at 0.1 mm^2 raster line of NIST 612 glass and monitoring the multi-element signal (*see* **Note 8**).
2. Two-dimensional distribution images are produced by running equally spaced, identical consecutive ablations parallel to one another to produce a raster image.
3. The spacing between the lines should be the diameter of the spot, so that the ablation lines do not cross.
4. Due to the restricted area on the stage within the sample chamber that the laser will ablate in, some of the tissues are cut in half to produce two sections and run separately. The two resulting images are then stitched together and should be clearly shown in the data with a white dotted line.
5. Prior to the analysis of the tissue of interest, a 60 s raster ablation of the reference materials with a 30 s gas blank period, before and after, is performed on the matrix-matched reference materials. This is also performed after the tissue is completed.

3.5.3 Limit of Detection (LOD) Calculation

1. Using an un-spiked reference material, run ten consecutive 60 s ablations, with 30 s gas blanks, before and after each run.
2. The limit of detection (LOD) for each element is calculated using software such as IOLITE 3.32 [9], based on Longerich et al. 1996, work, using the reference materials.
3. Alternatively, the LOD can be calculated manually using the following equation [10]:

$$LOD = \left[3 \times 1\sigma(\textit{background}) \times \surd(1/nb + 1/na)\right] / S$$

where

- 1σ(background) = 1 standard deviation of the background, measurement for this element.
- nb = number of measurements in the background integration.
- na = number of measurements in the sample integration.
- S = sensitivity.

4. An average is calculated from a minimum of eight replicates.

3.5.4 Calculation of Spatial Resolution of LA Images

The spatial resolution of an image is calculated for both the *X*-direction (the direction the laser passed) and the *Y*-direction (the directions the rasters followed). Figure 4 demonstrates that this is dependent on the direction in which the laser travels.

The *X*-direction is calculated as follows:

$$\text{Spatial resolution} = \text{SS} + (\text{DT} \quad \text{SR}) + \text{D washout}$$

- *SS* = Laser beam diameter.
- *DT* = Dwell time of a particular mass.

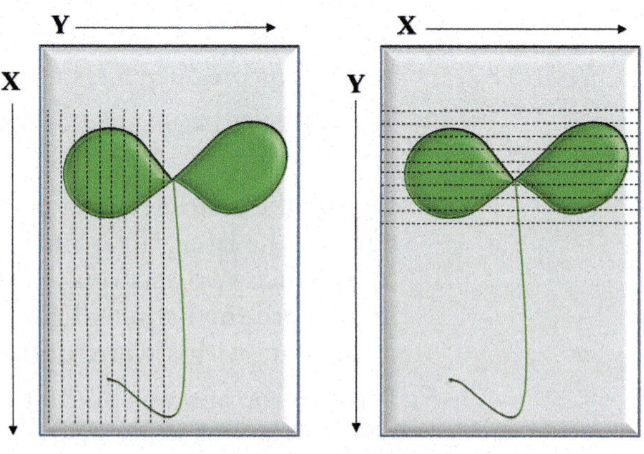

Fig. 4 Spatial resolution directions of tissues ablated differently

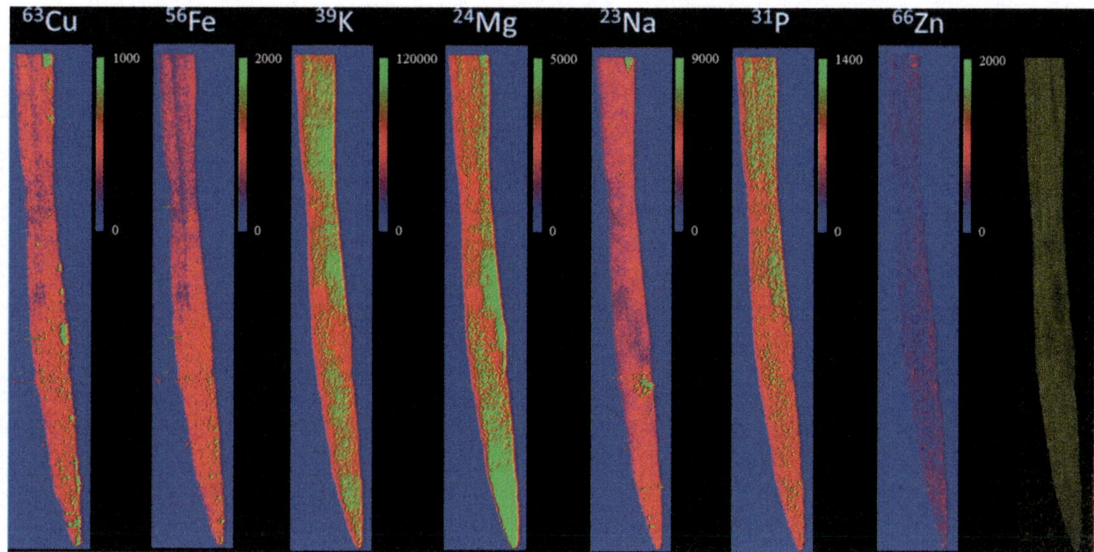

Fig. 5 2D elemental distribution images of a barley leaf

- *SR* = Scan speed/rate.
- *D washout* = sample chamber washout time × *SR*.

The sample chamber washout time is the time taken for the signal to go from 100% to 1% signal, once the laser has stopped firing. With the UP213 chamber this is approximately 20 s.

The spatial resolution in the *Y*-direction = the raster/scan line spacing.

This is usually the laser spot size, as the parallel raster lines are offset by the spot size in the *Y*-direction. Larger spacing is recommended to prevent contamination of adjacent tissue with debris from previous runs [11].

Figure 5 illustrates an image of a barley leaf when all of the parameters are optimal.

4 Notes

1. Gelatine-embedded material tends to produce better sections than CMC, as the gelatine produces slices that hold the material. On slicing CMC tends to disintegrate and also needs to remain below 0 °C to remain solid.
2. Thicker sections are more robust and are more suited for delicate material.
3. Carbon tape is high in sodium and, therefore, if analyzing for sodium, there could be an issue.

4. Dried seeds, such as barley, have a tendency not to thaw mount and require adhesion. Tissue with higher water content, such as radish bulbs, will easily thaw mount to a slide.
5. Be careful when using metal tools such as tweezers and scalpels as these can leave residues of metal behind, particularly zinc.
6. Reference material can be purchased from NIST (with a certificate of analysis), however, the type of material is limited and can be expensive.
7. All equipment is a potential source of contamination, so make sure all equipment is thoroughly cleaned in 5% nitric acid before use.
8. It is very important that the laser is in focus when it is passing over the tissue. Differences in the depth of the tissue should be accounted for by programming the laser height to change as it passes over the tissue. If the laser is not in focus, there is a dramatic drop in sensitivity.
9. When analyzing Se it is recommended to reduce the collision gas flow down to approximately 3.0 ml/min.

Acknowledgments

Rob Horton from ESI, Michael Cox SHU, Bence Paul and Dany Savard from IOLITE.

References

1. Jurowski K, Szewczyk M, Piekoszewski W et al (2014) A standard sample preparation and calibration procedure for imaging zinc and magnesium in rats' brain tissue by laser ablation-inductively coupled plasma-time of flight-mass spectrometry. J Anal At Spectrom 29(8):1425–1431. doi:10.1039/C3JA50378J
2. Hare D, Austin C, Doble P (2012) Quantification strategies for elemental imaging of biological samples using laser ablation-inductively coupled plasma-mass spectrometry. Analyst 137(7):1527–1537. doi:10.1039/c2an15792f
3. O'Connor C, Landon MR, Sharp BL (2007) Absorption coefficient modified pressed powders for calibration of laser ablation inductively coupled plasma mass spectrometry. J Anal At Spectrom 22(3):273–282. doi:10.1039/b612512c
4. Wu B, Zoriy M, Chen Y et al (2009) Imaging of nutrient elements in the leaves of Elsholtzia Splendens by laser ablation inductively coupled plasma mass spectrometry (LA-ICP-MS). Talanta 78(1):132–137. doi:10.1016/j.talanta.2008.10.061
5. Becker JS, Matusch A, Wu B (2014) Bioimaging mass spectrometry of trace elements – recent advance and applications of LA-ICP-MS: a review. Anal Chim Acta 835(0):1–18. http://dx.doi.org/10.1016/j.aca.2014.04.048
6. Cizdziel J, Bu K, Nowinski P (2012) Determination of elements in situ in green leaves by laser ablation ICP-MS using pressed reference materials for calibration. Anal Methods 4(2):564. doi:10.1039/C1AY05577A
7. Gomes MS, Schenk ER, Santos D Jr et al (2014) Laser ablation inductively coupled plasma optical emission spectrometry for analysis of pellets of plant materials. Spectrochim Acta B At Spectrosc 94–95(0):27–33. http://dx.doi.org/10.1016/j.sab.2014.03.005
8. Punshon T, Jackson B, Bertsch P et al (2004) Mass loading of nickel and uranium on plant surfaces: application of laser ablation-ICP-MS. J Environ Monit 6(2):153–159. doi:10.1039/b310878c

9. Paton C, Hellstrom J, Paul B et al (2011) Iolite: freeware for the visualisation and processing of mass spectrometric data. J Anal At Spectrom 26(12):2508. doi:10.1039/C1JA10172B

10. Longerich HP, Jackson SE, Gunther D (1996) Inter-laboratory note. Laser ablation inductively coupled plasma mass spectrometric transient signal data acquisition and analyte concentration calculation. J Anal At Spectrom 11(9):899–904. doi:10.1039/JA9961100899

11. Hutchinson RW, Cox AG, McLeod CW et al (2005) Imaging and spatial distribution of β-amyloid peptide and metal ions in Alzheimer's plaques by laser ablation–inductively coupled plasma–mass spectrometry. Anal Biochem 346(2):225–233. http://dx.doi.org/10.1016/j.ab.2005.08.024

Chapter 12

Mass Spectrometry Imaging of Drugs of Abuse in Hair

Bryn Flinders, Eva Cuypers, Tiffany Porta, Emmanuel Varesio, Gérard Hopfgartner, and Ron M.A. Heeren

Abstract

Hair testing is a powerful tool routinely used for the detection of drugs of abuse. The analysis of hair is highly advantageous as it can provide prolonged drug detectability versus that in biological fluids and chronological information about drug intake based on the average growth of hair. However, current methodology requires large amounts of hair samples and involves complex time-consuming sample preparation followed by gas or liquid chromatography coupled with mass spectrometry. Mass spectrometry imaging is increasingly being used for the analysis of single hair samples, as it provides more accurate and visual chronological information in single hair samples.

Here, two methods for the preparation of single hair samples for mass spectrometry imaging are presented.

The first uses an in-house built cutting apparatus to prepare longitudinal sections, the second is a method for embedding and cryo-sectioning hair samples in order to prepare cross-sections all along the hair sample.

Key words MALDI-MSI, MetA-SIMS, Cocaine, Longitudinal sectioning, Cross-section

1 Introduction

Hair testing is a powerful tool routinely used for the detection of drugs of abuse in toxicology and forensic applications [1–3]. The analysis of hair is highly advantageous as it can provide prolonged detection and chronological information about drug intake or chemical exposure in contrast to the analysis of biological fluids [4]. However, current methodology requires large amounts of hair samples and involves complex and time-consuming sample preparation, which includes homogenization, derivatization, sample-cleanup, and extraction techniques followed by gas or liquid chromatography coupled with mass spectrometry (GC-MS or LC-MS).

The use of matrix-assisted laser desorption/ionization-mass spectrometry imaging (MALDI-MSI) for the detection and imaging of drugs and pharmaceuticals in tissues is well established.

Laura M. Cole (ed.), *Imaging Mass Spectrometry: Methods and Protocols*, Methods in Molecular Biology, vol. 1618,
DOI 10.1007/978-1-4939-7051-3_12,

However, it is increasingly being used for the analysis of drugs of abuse in hair. The feasibility of using matrix-assisted laser desorption/ionization-mass spectrometry (MALDI-MS) for the analysis of drug of abuse in hair was initially explored in order to detect cocaine and its metabolites in hair, following pulverization and extraction [5, 6]. Following this, the detection and imaging of cocaine and its metabolites in single intact user hair samples were demonstrated using a MALDI-triple quadrupole linear ion trap instrument [7]. Some preliminary results on the determination of cocaine and cannabinoids in single intact hair samples using a MALDI-LTQ Orbitrap XL instrument have also been presented [8]. MALDI-tandem mass spectrometry (MALDI-MS/MS) imaging has been used to monitor the distribution of pharmaceuticals, such as the synthetic opioid painkiller tilidine in intact hair samples of children obtained from a forensic intoxication case [9]. The detection of zolpidem in single intact hair samples was performed by MALDI-MS and MALDI-Fourier transform ion cyclotron resonance mass spectrometry (FTICR-MS) imaging [10]. The effect of hydrogen peroxide treatment on both cocaine users and externally contaminated hair samples has also been investigated using MALDI-MSI [11].

While the analysis of intact hair samples has been demonstrated by MALDI-MS imaging, one of the issues often raised is the extraction efficiency of the embedded drugs by the matrix solution. As the drugs are considered to be bound to melanin inside the core of the hair, it remains difficult to know whether the drug is completely extracted out of the hair by the MALDI matrix especially through the impermeable outer surface. The preparation of longitudinal sections has been proposed as a solution to address this issue. This also opens the possibility of analyzing drug user hair samples with surface analysis techniques such as secondary ion mass spectrometry (SIMS).

MALDI-MS and MALDI-FTICR-MS imaging were used to measure methamphetamine in longitudinally sectioned hair samples [12, 13]. MALDI-FTICR-MS imaging was also used to monitor the distribution of ketamine in longitudinal scraped scalp hairs from chronic users [14]. MALDI-MS/MS imaging has also been used for the detection of nicotine in longitudinally sectioned hair from heavy smokers [15].

Recently, MALDI-MS/MS imaging was employed to investigate the incorporation of methoxyphenamine (an analogue of methamphetamine) into hair. The analysis was performed at various time points on single longitudinal sectioned hairs from volunteers after a single oral administration [16]. The consequences of different washing procedures on the distribution of cocaine in hair were recently investigated using MALDI-MS/MS and metal-assisted secondary ion mass spectrometry (MetA-SIMS) imaging [17].

In this chapter, we present two methods for the preparation of longitudinal and cross-sections of single hair samples for the analysis of drugs of abuse by MALDI-MS/MS and MetA-SIMS imaging.

2 Materials

2.1 Chemicals and Materials

1. Gelatin from porcine skin.
2. Double-sided copper tape.
3. Double-sided tape.
4. Methanol (MeOH).
5. Dichloromethane (DCM).
6. Trifluroacetic acid (TFA)
7. α-Cyano-4-hydroxycinnamic acid (CHCA) matrix solution (10 mg/mL in 70:30 MeOH: H_2O with 0.2% TFA).
8. Cocaine base.
9. Indium tin oxide (ITO)-coated glass slides, 25 × 50 × 1.1 mm, 4–8 Ω resistance
10. Small magnets.
11. Plastic tubes.

2.2 Instruments

1. Cryo-microtome.
2. Bruker ImagePrep matrix application device.
3. Leica DMRX light microscope equipped with a digital camera.
4. Quorum Technologies sputter coater equipped with a gold target, quartz crystal microbalance, and film thickness monitor.
5. Waters MALDI HDMS Synapt mass spectrometer equipped with a 200 Hz, 335 nm neodymium-doped yttrium aluminum garnet (Nd:YAG) laser.
6. Physical Electronics TRIFT II TOF-SIMS mass spectrometer equipped with an Au liquid metal ion gun tuned for 22 keV Au^+ primary ions.

2.3 Software

1. BioMap 3.7.5.5 software.
2. Waters MALDI imaging pattern creator software.
3. Waters MALDI imaging converter software.
4. Waters MassLynx 4.1 software.
5. Physical Electronics WinCadence 4.4.0.17 software.
6. Physical Electronics Vacuum Watcher 3.1.4.3 software.

3 Methods

3.1 Sample Decontamination

1. Intact single hair samples were placed into a glass vial containing 10 mL of dichloromethane and shaken for 1 min (*see* **Notes 1** and **2**). Retain the wash solutions for further analysis, if required.
2. The hair samples were left to dry at room temperature before further preparation.

3.2 Longitudinal Sectioning

1. The in-house built cutting apparatus shown in Fig. 1a consists of a stainless-steel block (60 × 110 × 10 mm) with 20–80 μm grooves spaced 5 mm apart, which is based on previously reported specifications [18]. In addition to this is the cutting device, which is also made from stainless steel and holds half of a cryo-microtome blade at a 30° angle [19]. The schematics of the cutting apparatus are shown in Fig. 1b.
2. Following decontamination, place a single hair sample into one of the grooves of the cutting block and fix at one end with a small piece of double-sided tape, as shown in Fig. 2a (*see* **Notes 3–5**).
3. While holding the other end with a gloved finger, slowly run the cutting device along the length of the hair (Fig. 2b).
4. Place another piece of double-sided tape on the other end of the hair and transfer onto double-sided copper tape mounted on a ITO glass slide, press the hair into the tape using a clean glass slide (Fig. 2c).
5. Use a Leica DMRX microscope equipped with a digital camera to determine if the hair section has been successfully transferred onto the glass slide.

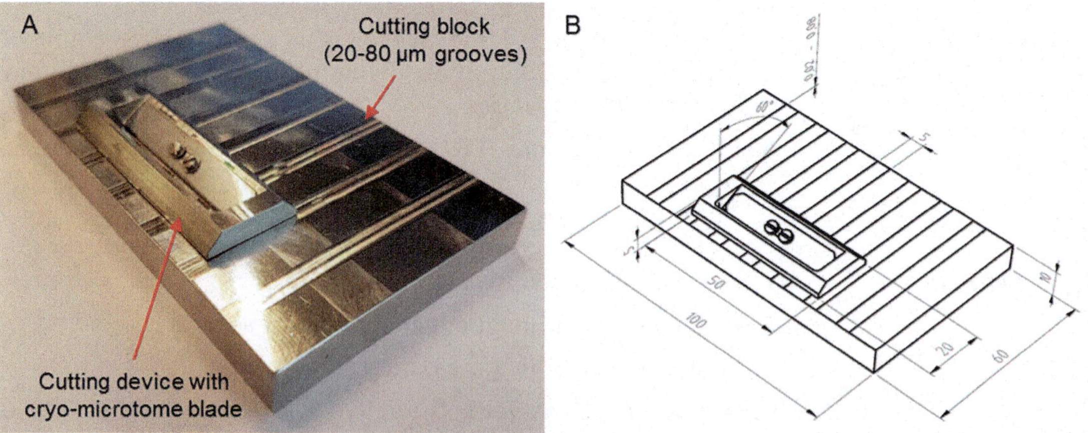

Fig. 1 Cutting apparatus used for longitudinal sectioning of hair samples. (**a**) Image of the in-house built cutting apparatus and (**b**) schematic of the cutting apparatus

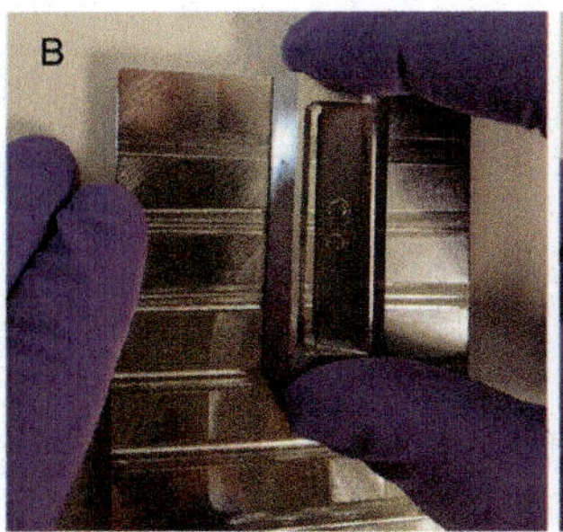

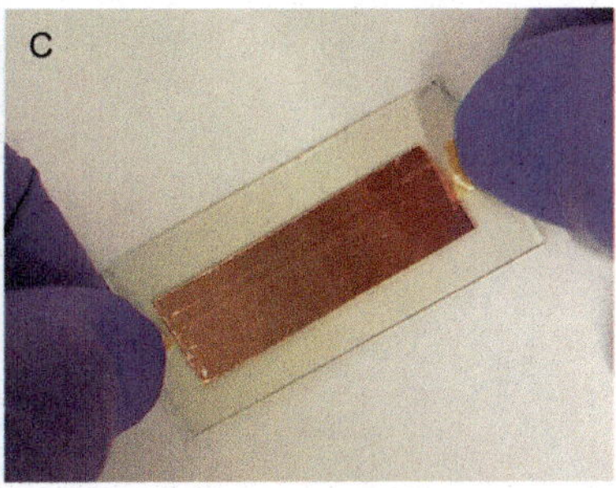

Fig. 2 Method for preparing longitudinal sections of single hair samples. (**a**) Hair sample attached to the cutting block with double-sided tape, (**b**) preparing longitudinal sections of single hair samples using in-house built cutting apparatus, (**c**) longitudinal sections mounted on conductive glass slide with double-sided copper tape

3.3 Preparation of Cross-Sections

1. Following decontamination, thread the hair sample through the lid of the tube and using a paper clip, attach two small magnets at the end of the single hair sample. Attach two more magnets at the opposite end of the hair sample.
2. Place the hair sample into a plastic tube that contains a 10% solution (w/v) of gelatin, as shown in Fig. 3b (*see* **Note 6**).
3. Snap-freeze the contents by placing the plastic tube in liquid nitrogen for 30 s (Fig. 3c).
4. Remove the embedded hair from the plastic tube and cut into 1 cm blocks (the average growth of hair being 1 cm/month), mount one of the blocks onto a cryo-microtome stage using a drop of water.
5. Section the embedded hair samples at −20 °C using a cryo-microtome to produce 12 μm thick sections, thaw mount sections onto a clean indium tin oxide (ITO) glass slide (*see* **Note 7**).
6. Inspect the sections using a Leica DMRX microscope equipped with a digital camera to determine the position of the hair cross-section and mark the location on the opposite side of the glass slide using a permanent marker pen.

3.4 Matrix Deposition for MALDI-MS/MS Imaging

1. Place the sample in the ImagePrep matrix application device.
2. Fill the bottle with the matrix solution.
3. Select the manufacturer's method for spraying CHCA and start the sequence.
4. Use a Leica DMRX microscope equipped with a digital camera to inspect the crystal coverage.

3.5 Metal Deposition for MetA-SIMS Imaging

1. Place the sample in the chamber of the Quorum Technologies sputter coater and close the lid.
2. Select the density of the metal (19.30 g/cm^3 in the case of gold) and the desired thickness (1 nm) on the film thickness monitor.

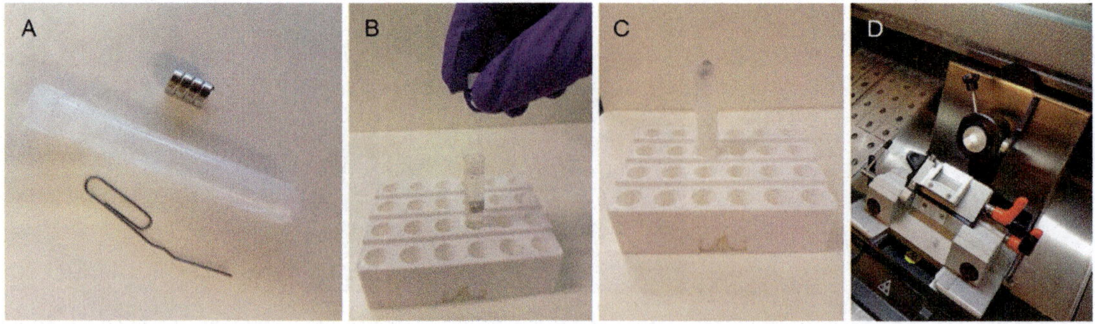

Fig. 3 Method for embedding single hair samples. (**a**) Materials required for the embedding of single intact hair samples, (**b**) single hair sample clipped between magnets and placed in a plastic tube filled with embedding material, (**c**) embedded hair sample following snap freezing, and (**d**) cryo-sectioning of embedded hair samples

3. Set the discharge voltage to 1.5 kV and the plasma current to 25 mA in order to achieve a homogenous coating.
4. Start the sequence (operate in automatic mode), once the sequence is complete vent the instrument and remove the sample (*see* **Note 8**).

3.6 MALDI-MS/MS Imaging

1. Calibrate the instrument prior to analysis with either a standard mixture of polyethylene glycol (PEG 200-3000) in water mixed with matrix or a saturated solution of red phosphorus in acetone.
2. Spot 0.5 μL of a cocaine base standard (100 ng/μL in 70% MeOH) onto a MALDI target plate, followed by 0.5 μL of the matrix solution.
3. Optimize the instrumental settings using the cocaine base standard such as the laser power and collision energy (trap cell). The optimal settings were as follows: laser power 250 (200 Hz) and collision energy 10 eV (monitor the main product ion of cocaine at m/z 182 formed by neutral loss of benzoic acid).
4. Following method optimization, attach the sample onto a glass slide adaptor using double-sided tape and scan using a flatbed scanner to produce a digital image, ensure the image quality is around 600 dpi or better.
5. Import the digital image of the sample into the MALDI imaging pattern creator software, to define the area to be imaged and the spatial resolution (150 × 50 μm).
6. Set up the MS/MS imaging method in the MassLynx 4.1 software. The settings were as follows: positive ion mode, V-mode, mass range m/z 50–350, also use the previously optimized settings (*see* **step 3**).

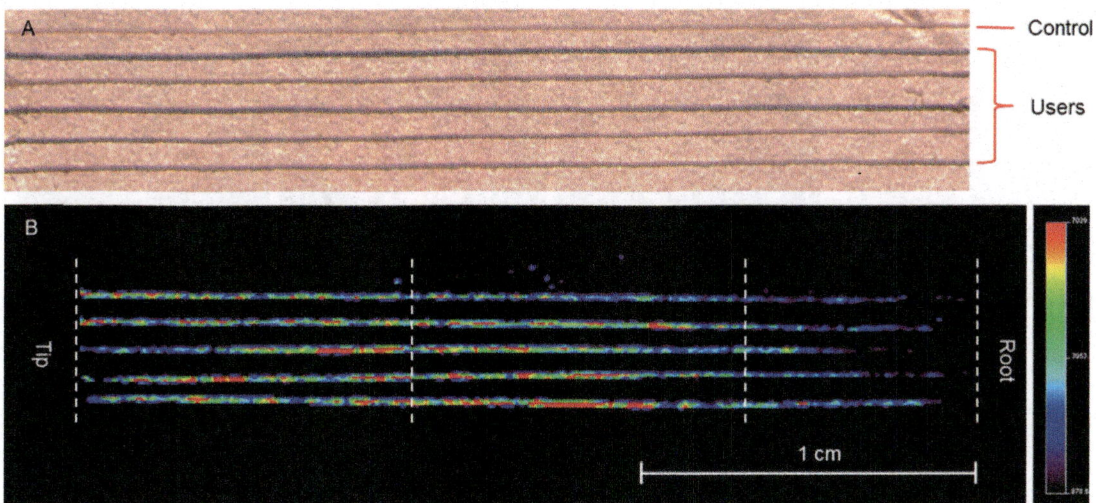

Fig. 4 MALDI-MS/MS imaging of longitudinal sectioned drug user hair. (**a**) Optical image of longitudinal sectioned control and drug user hair samples mounted on double-sided copper tape and (**b**) MALDI-MS/MS image showing the distribution of the cocaine product ion at m/z 182. The length of the hair samples analyzed was 2.7 cm, given that the average growth rate of human hair is around 1 cm per month this corresponds to a growth period of approximately 3 months. Since the spatial resolution along the hair is 150 μm, each pixel is equivalent to around 12 h of growth (reproduced from ref. 19 with permission from John Wiley & Sons, Ltd)

7. Following data acquisition, convert the raw data (.raw) into the Analyze file format (.img) using the MALDI imaging converter software.
8. Visualize the converted data using the BioMap 3.7.5.5 software (Fig. 4).

3.7 MetA-SIMS Imaging

1. Place the sample in the target holder and scan using a flat-bed scanner to produce a digital image, import the image into the stage control window.
2. Load the sample into the instrument and locate the sample using the scan in the stage control window and fine tune with the joy-stick.
3. Use a single 100 × 100 μm tile (containing 256 × 256 pixels) for imaging of hair cross-sections, images were acquired in positive ion mode for 15 min.
4. Use the mosaic mapping function to image a portion of the longitudinally sectioned hair sample. Images were acquired on a 2 mm portion of the longitudinally sectioned hair sample, which consisted of 32 × 7 tiles (each tile was 100 × 100 μm containing 256 × 256 pixels).
5. The results are calibrated using elements such as sodium (*m/z* 23), potassium (*m/z* 39), indium (*m/z* 115), or gold (*m/z* 197). Common hydrocarbon fragments such as CH_3^+ (*m/z* 15), $C_2H_3^+$ (*m/z* 27), and $C_4H_7^+$ (*m/z* 55) can also be used.

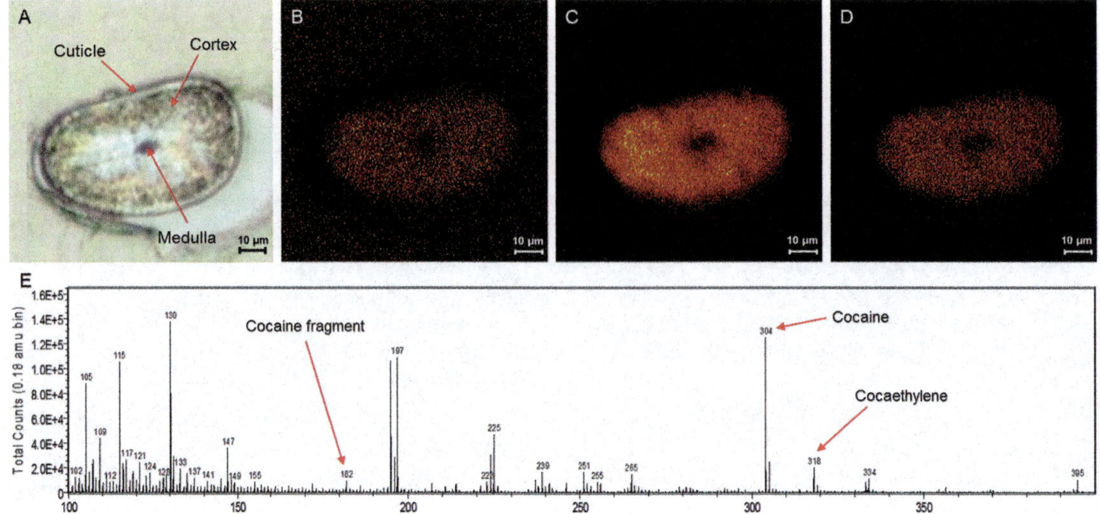

Fig. 5 MetA-SIMS imaging of cross-sections of drug user hair. (**a**) Optical image of hair cross section, (**b**) MetA-SIMS images showing the distribution of a cocaine fragment at m/z 182, (**c**) cocaine at m/z 304, (**d**) cocaethylene at m/z 318, and (**e**) average positive ion spectrum from drug users hair. The thickness of the cross-section was 12 μm, given that the average growth rate of human hair is around 1 cm per month this corresponds to a growth period of around 1 h

6. Select peaks of interest and visualize using the WinCadance 4.4.0.17 software (Figs. 5 and 6).

4 Notes

1. Decontamination of hair samples prior to analysis is an important step that serves two purposes. The first is to remove hair products, sweat, sebum, and other surface materials that can interfere with the analysis. The second is to remove any drug material deposited on the hair from sweat or environmental contamination [20].
2. The Society of Hair Testing (SOHT) currently recommends the use of organic solvents such as acetone or dichloromethane for decontamination, as aqueous solutions or methanol can cause the hair to swell causing extraction or incorporation of external contamination [20]. However, various laboratories and law enforcement agencies will have their own validated decontamination procedures.
3. Approximately 0.3 cm of the hair is covered with double-sided tape to fix the hair onto the cutting block and to transfer to the glass slide.
4. The minimum length of hair that can be prepared with the cutting apparatus is 1 cm, this is due to the maneuvering of the cutting device.

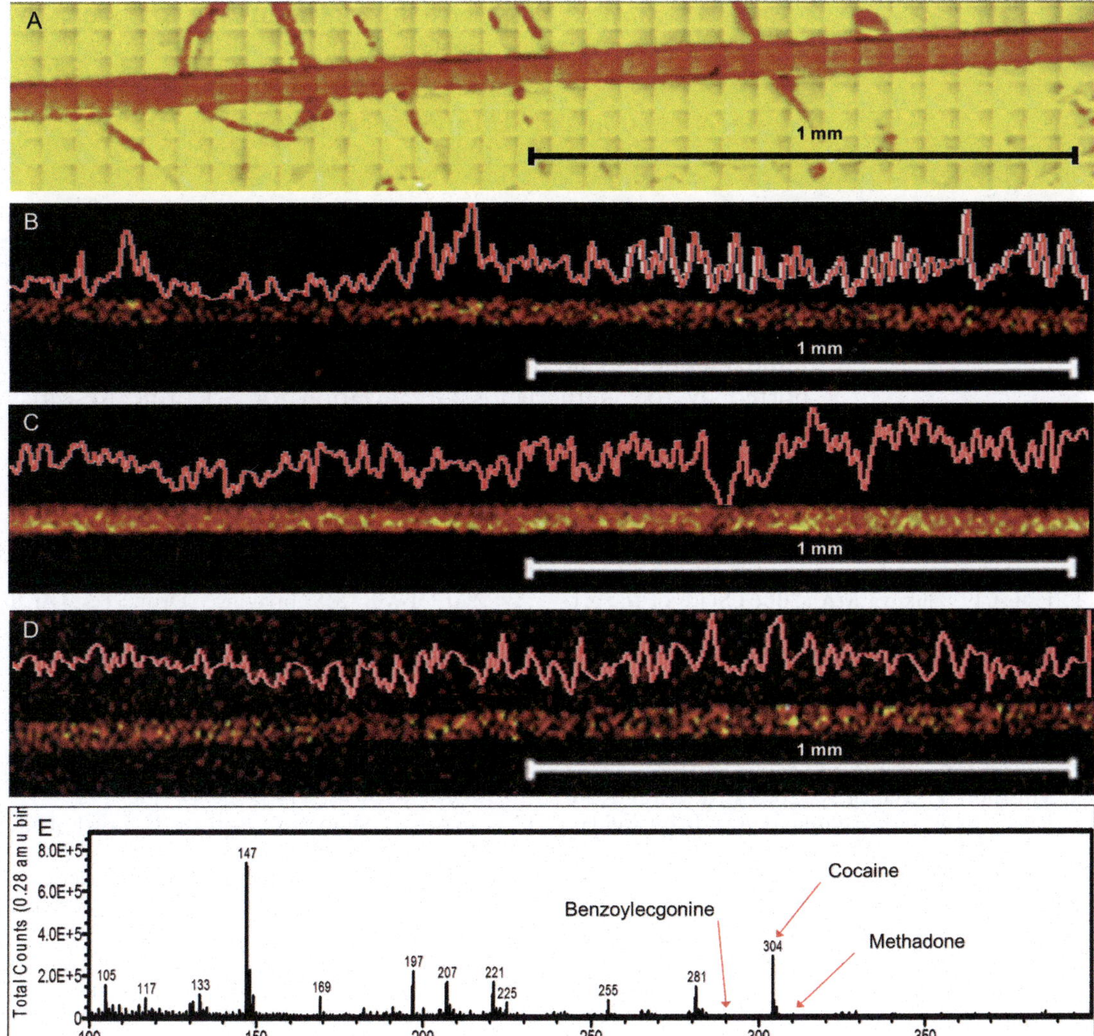

Fig. 6 MetA-SIMS imaging of longitudinal sectioned drug user hair. MetA-SIMS images of the (**a**) total ion current, (**b**) distribution of benzoylecgonine at m/z 290, (**c**) cocaine at m/z 304, (**d**) methadone at m/z 310, and (**e**) average positive ion spectrum from drug users hair. The area of the analyzed sample was 2 mm, given that the average growth rate of human hair is around 1 cm per month this corresponds to a growth period of 6 days. The smallest pixel size that can be achieved is 1 μm, which is theoretically equivalent to around 5 min of hair growth (reproduced from ref. 19 with permission from John Wiley & Sons, Ltd)

5. The maximum length of hair that can be prepared with the cutting apparatus is 6 cm, longer hairs need to be cut into segments and sectioned individually.
6. A 1% or 2% (w/v) solution of carboxymethylcellulose (CMC) or a 10% (w/v) solution of trehalose was also found to be suitable materials for embedding single intact hair samples.
7. As the ITO slides come wrapped in plastic it is necessary to clean them with hexane and ethanol, as polydimethylsiloxane

(PDMS)-related peaks can hinder the analysis of compounds in the low mass range with TOF-SIMS.

8. Gold coating has previously been shown to improve the secondary ion yield in SIMS, as well as to reduce surface charging on insulating samples [21].

Acknowledgments

This work is part of the research program of the Foundation of Fundamental Research on Matter (FOM) which is financially supported by the Netherlands Organization for Scientific Research (NWO). Part of this work was funded by the NWO Forensic Science program (project nr 727.011.004).

References

1. Pragst F, Balikova MA (2006) State of the art in hair analysis for detection of drug and alcohol abuse. Clin Chim Acta 370:17–49
2. Vincenti M, Salomone A, Gerace E, Pirro V (2013) Application of mass spectrometry to hair analysis for forensic toxicological investigations. Mass Spectrom Rev 32:312–332
3. Musshoff F, Madea B (2007) New trends in hair analysis and scientific demands on validation and technical notes. Forensic Sci Int 165:204–215
4. Kintz P, Villain M, Cirimele V (2006) Hair analysis for drug detection. Ther Drug Monit 28:442–446
5. Vogliardi S, Favretto D, Frison G, Ferrara SD, Seraglia R, Traldi P (2008) A fast screening MALDI method for the detection of cocaine and its metabolites in hair. J Mass Spectrom 44:18–24
6. Vogliardi S, Favretto D, Frison G, Maietti S, Viel G, Seraglia R, Traldi P, Ferrara SD (2010) Validation of a fast screening method for the detection of cocaine in hair by MALDI-MS. Anal Bioanal Chem 396:2435–2440
7. Porta T, Grivet C, Kraemer T, Varesio E, Hopfgartner G (2011) Single hair cocaine consumption monitoring by mass spectrometric imaging. Anal Chem 83:4266–4272
8. Musshoff F, Arrey T, Strupat K (2013) Determination of cocaine, cocaine metabolites and cannabinoids in single hairs by MALDI Fourier transform mass spectrometry-preliminary results. Drug Test Anal 5:361–365
9. Poetzsch M, Baumgartner MR, Steuer AE, Kraemer T (2015) Segmental hair analysis for differentiation of tilidine intake from external contamination using LC-ESI-MS/MS and MALDI-MS/MS imaging. Drug Test Anal 7:143–149
10. Poetzsch M, Steuer AE, Roemmelt AT, Baumgartner MR, Kraemer T (2014) Single hair analysis of small molecules using MALDI-triple quadrupole MS imaging and LC-MS/MS: investigations on opportunities and pitfalls. Anal Chem 86:11758–11765
11. Cuypers E, Flinders B, Bosman IJ, Lusthof KJ, Van Asten AC, Tytgat J, Heeren RMA (2014) Hydrogen peroxide reactions on cocaine in hair using imaging mass spectrometry. Forensic Sci Int 242:103–110
12. Miki A, Katagi M, Kamata T, Zaitsu K, Tatsuno M, Nakanishi T, Tsuchihashi H, Takubo T, Suzuki K (2011) MALDI-TOF and MALDI-FTICR imaging mass spectrometry of methamphetamine incorporated in hair. J Mass Spectrom 46:411–416
13. Miki A, Katagi M, Shima N, Kamata H, Tatsuno M, Nakanishi T, Tsuchihashi H, Takubo T, Suzuki K (2011) Imaging of methamphetamine incorporated into hair by MALDI-TOF mass spectrometry. Forensic Toxicol 29:111–116
14. Shen M, Xiang P, Shi Y, Pu H, Yan H, Shen B (2014) Mass imaging of ketamine in a single scalp hair by MALDI-FTMS. Anal Bioanal Chem 406:4611–4616
15. Nakanishi T, Nirasawa T, Takubo T (2014) Quantitative mass barcode-like image of nicotine in single longitudinally sliced hair sections from long-term smokers by matrix-assisted laser desorption time-of-flight mass spectrometry imaging. J Anal Toxicol 38:349–353
16. Kamata T, Shima N, Sasaki K, Matsuta S, Takei S, Katagi M, Miki A, Zaitsu K, Nakanishi T, Sato T, Suzuki K, Tsuchihashi H (2015) Time-course

mass spectrometry imaging for depicting drug incorporation into hair. Anal Chem 87: 5476–5481

17. Cuypers E, Flinders B, Boone CM, Bosman IJ, Lusthof KJ, Van Asten AC, Tytgat J, Heeren RMA (2016) Consequences of decontamination procedures in forensic hair analysis using metal-assisted secondary ion mass spectrometry analysis. Anal Chem 88:3091–3097
18. Kempson IM, Skinner WM, Kirkbride PK (2002) A method for the longitudinal sectioning of single hair samples. J Forensic Sci 47: 889–892
19. Flinders B, Cuypers E, Zeijlemaker H, Tytgat J, Heeren RMA (2015) Preparation of longitudinal sections of hair samples for the analysis of cocaine by MALDI-MS/MS and TOF-SIMS imaging. Drug Test Anal 7:859–865
20. Cooper GAA, Kronstrand R, Kintz P (2012) Society of hair testing guidelines for drug testing in hair. Forensic Sci Int 218:20–24
21. Altelaar AFM, Klinkert I, Jalink K, de Lange RPJ, Adan RAH, Heeren RMA, Piersma SR (2006) Gold-enhanced biomolecular surface imaging of cells and tissue by SIMS and MALDI mass spectrometry. Anal Chem 78:734–742

Chapter 13

MALDI Mass Spectrometry Profiling and Imaging Applied to the Analysis of Latent Fingermarks

Robert Bradshaw

Abstract

Latent fingermarks are derived from a transfer of material from the fingertips to a surface upon contact. Traditionally, fingermarks are employed for biometric identification of individuals based on matching of the pattern of the ridges. However, in recent years, there has been a stark increase in the use of advanced analytical techniques in order to obtain additional information, specifically the chemical composition of the residue. Understanding the complexity of the endogenous and exogenous content of fingermarks could be extremely useful in allowing further development of enhancement techniques currently used in forensic scenarios by identifying potential target molecules. This chemical information could also potentially provide invaluable information on the lifestyle of an individual, including their activities prior to depositing a mark.

An analytical tool that has gained notable popularity in this novel area of research is matrix-assisted laser desorption/ionisation mass spectrometry (MALDI MS). This technique can either be employed for rapid chemical *profiling* or *imaging* of fingermarks to detect chemical species contained within the residue, with the latter also allowing for physical reconstruction of the fingermark ridges.

This chapter will provide an overview of the protocols employed to allow for both MALDI MS *profiling* and *imaging* analysis of latent fingermarks, specifically covering the types of fingermarks employed and techniques used to deposit matrices onto samples.

Key words Fingermark, Forensic, MALDI MS, Mass spectrometry Imaging, Matrix

1 Introduction

Fingermarks are derived from a transfer of material from the friction ridge skin of our fingertips to a surface upon contact. The general pattern of the ridges and small characteristics, known as *minutiae*, cannot be duplicated among individuals [1]. This has enabled the use of fingermarks as an effective means of biometric identification for more than a century [2].

Over recent years, the forensic and analytical community has gone to great lengths to investigate the chemical composition of latent fingermark residue using modern analytical techniques [3]. Extending our knowledge of the endogenous composition of fingermarks could enable further development of fingermark enhancement

Laura M. Cole (ed.), *Imaging Mass Spectrometry: Methods and Protocols*, Methods in Molecular Biology, vol. 1618,
DOI 10.1007/978-1-4939-7051-3_13, © Springer Science+Business Media LLC 2017

processes. Furthermore, the ability to identify exogenous contaminants contained within fingermark residue can deliver invaluable intelligence about what a person may have touched or even ingested before excreting specific molecules (such as metabolites) through sweat.

Research into fingermark composition has been conducted using a wide range of novel analytical techniques, including immunodetection [4–11], spectroscopy [12–20], and mass spectrometry. The latter has been studied extensively in various formats, including gas chromatography-mass spectrometry (GC-MS) [3, 21–27], liquid chromatography-mass spectrometry (LC-MS) [10, 28], desorption electrospray ionisation mass spectrometry (DESI-MS) [29–31], surface-assisted laser desorption/ionisation-mass spectrometry (SALDI-MS) [32–39], gold nanoparticle-assisted mass spectrometry [40], secondary ion mass spectrometry (SIMS) [41–47], direct analysis in real time-mass spectrometry (DART-MS) [37, 48], and laser desorption ionization-mass spectrometry (LDI-MS) [49, 50].

One particular area of mass spectrometry that has remained at the forefront of this research is matrix-assisted laser desorption ionization mass spectrometry (MALDI MS) profiling and imaging [51]. Since the initial development of the use of this technique for the analysis of latent fingermarks [52], a range of applications have now been developed, including the identification of condom lubricants that have been used to contaminate marks (with potential uses in cases of sexual assault) [18, 53], separation of overlapping fingermarks [54], the confirmation of blood in marks [55–57], detection of drugs and excreted metabolites [58, 59], the detection of explosives [60], prediction of sex from statistical analysis of excreted peptides [61], and the development of forensically applicable workflows [62–64]. A recently published review provides an extensive overview of the application of MALDI MS to fingermark analysis [51].

Besides the benefits of analyzing latent fingermarks by MALDI MS as a means to obtain forensic intelligence, this sample type is an excellent choice for routine method development; preparation of fingermarks takes a maximum of 15 min, they are in unlimited supply and can act as a homogenous, relatively reproducible sample type. This chapter aims to provide an insight into the methodologies that have been optimized for the MALDI MS analysis of fingermarks in a number of scenarios.

2 Materials

2.1 Preparation of Latent Fingermarks

1. Fingermark substrate: ALUGRAM1 SIL G/UV254 precoated aluminum sheets.
2. Surface/hand wash—200 mL of 50:50 EtOH:Deionised water.
3. Lint-free absorbent tissue.

4. Plastic bag or powder-free nitryl gloves [2] for the preparation of eccrine marks.
5. Sterile lancet for the preparation of blood contaminated marks.
6. Glass microscope slides for the preparation of exogenous substance contaminated marks.
7. TAAB Double-sided carbon tape.
8. Klenair air spray (Kenro Ltd).
9. Pair of scissors with metal blades.

2.2 Matrix Deposition onto Latent Fingermarks

1. Typical matrices (*see* **Note 1**).

   α-Cyano-4-hydroxycinnamic acid (α-CHCA).

   2,5-dihydroxybenzoic acid (DHB).

   Sinapinic acid (SA).

   Dithranol.

   Curcumin.

   9-Aminoacridine (9-AA).

   2-mercaptobenzothiazole (MBT).

2.3 Typical Solvents Required

Acetonitrile (ACN).

0.1–0.5% trifluoroacetic acid (TFA) in deionized water.

Ethanol (EtOH).

Methanol (MeOH).

Dichloromethane (DCM).

Tetrahydrofuran (THF).

Acetone.

3 Methods

3.1 Substrate Preparation

In a forensic setting, fingermarks can be discovered on a wide range of different surfaces; however in an experimental setting, certain substrates are more advantageous for subsequent MALDI MS analysis. ALUGRAM1 SIL G/UV254 precoated aluminum sheets have been used extensively in the analysis of latent fingermarks [18, 52–58, 61–66] (*see* **Note 2**).

1. Prepare a wash bottle containing 100% acetone.
2. Place the TLC plate onto a flat surface with the (white) silica coating facing upward.
3. Deposit approximately 10 mL of acetone directly onto the silica.
4. Using absorbent tissue, gently rub the surface until the silica coating is removed. Avoid excessive manipulation of the aluminum sheets as they may bend.

5. Repeat **steps 3** and **4** until the aluminum plate is devoid of any silica.
6. Clean both sides of the TLC plate thoroughly using approximately 10 mL of acetone, ensuring that the surface remains against a flat surface to avoid bending or dimpling of the surface.
7. Clean the blades of a pair of metal scissors thoroughly with acetone and absorbent tissue paper.
8. Cut the aluminum to the desired size.
9. Fingermarks should be deposited onto the side of the aluminum plate that was not coated in silica.
10. Once the fingermark has been prepared, the substrate can be attached to the MALDI target plate using a double-sided conductive carbon tape.
11. Use the end point of a scalpel to ensure that the aluminum plate is completely secured to the MALDI target plate by pushing down around the entirety of the perimeter at 5 mm increments.

3.2 Fingermark Deposition

A number of different types of fingermarks can be prepared depending on the outcomes required from a specific experiment, namely; ungroomed, groomed, eccrine, natural, bloodied, and contaminated. It is imperative to control certain conditions when depositing fingermarks to minimize variability (*see* **Notes 3–5**).

3.2.1 Ungroomed Fingermarks

Ungroomed fingermarks are the most reproducible type of marks and should be used when conducting experiments over extended periods of time. This sample type contains the minimal amount of fingermark material and can be used to assess the sensitivity of a particular methodology. The protocol outlined below was first employed in the initial MALDI MS study of latent fingermarks [52] and has been used extensively in subsequent research in this area.

1. Using the surface/hand wash (50:50 EtOH: Deionized water) and absorbent tissue, clean surfaces around a computer workstation, including the keyboard and mouse. Ensure that any residual moisture is removed.
2. Clean both hands thoroughly using approximately 3 mL of the hand wash solution and dry thoroughly using absorbent tissue.
3. Conduct computer activities for a total of 15 min, ensuring that both hands only come to contact with areas that have been cleaned to avoid contamination of the fingertips.
4. After the allocated time, rub the fingertips from both hands together to ensure distribution of excreted components across the fingertips.
5. Deposit a fingermark onto the desired surface.

3.2.2 Groomed (Sebaceous) Fingermarks

Groomed fingermarks are employed when a sebum-rich mark containing an abundance of chemical content is desired, this is often required when performing initial tests on the effectiveness of a methodology [2].

1. Clean both hands thoroughly using approximately 3 mL of the surface/hand wash solution and dry thoroughly using absorbent tissue.
2. Rub the fingertips of both hands numerous times across the bridge of your nose.
3. Rub the fingertips from both hands together to distribute the material evenly.
4. Deposit a mark onto the desired surface.

3.2.3 Natural Fingermarks

Natural fingermarks are a representation of what to expect in a "realistic" forensic scenario, as no fingermark pre-preparation (cleaning or grooming) has occurred. Therefore, this type of mark may contain a wide range of endogenous/exogenous species and is likely to exhibit the most amount of variation.

1. Ensure that the hands have not been cleaned at least 1 h prior to the deposition of a natural mark, normal daily activities should be conducted during this time.
2. Rub the fingertips of both hands together to ensure an even distribution of fingermark components.
3. Deposit a mark onto the desired surface.

3.2.4 Eccrine Fingermarks

Eccrine fingermarks are useful for studying the amino acid/protein content of sweat in fingermarks [2] and have previously been employed to analyze endogenous peptides contained within fingermark residue by MALDI MS [61].

1. Clean both hands thoroughly using approximately 3 mL of the hand wash solution and dry thoroughly using absorbent tissue.
2. Place one hand in a plastic bag/powder-free nitryl gloves and secure around the wrist using a piece of string/elastic band.
3. Keep the hand inside the bag for 15 min.
4. Remove the hand from the bag and rub the fingertips of this hand together.
5. Deposit a mark onto the desired surface.

3.2.5 Exogenous Substance Contaminated Fingermarks

1. Clean a glass slide thoroughly using a sequence of 1 mL ethanol followed by 1 mL acetone, ensuring that the glass slide is completely dry by wiping with lint-free absorbent tissue.
2. Prepare a solution of the exogenous substance of interest (e.g., 10 μg/mL cocaine in 100% MeOH) (*see* **Note 6**).

3. Follow previous instructions depending on the type of fingermark required (ungroomed/groomed/natural/eccrine).
4. Arrange for a colleague to deposit 100 μL of the exogenous compound solution across the entirety of the glass slide surface using an automatic pipette.
5. Once the solvent has completely evaporated, rub a fingertip or an artificial finger (*see* **Note 7**) thoroughly across the glass slide in order to transfer the exogenous substance residue to the fingertip.
6. Deposit a mark onto the desired surface.
7. Clean the hands thoroughly using soap and water to remove any remnants of the exogenous substance from the fingertip.

3.2.6 Bloodied Fingermarks

Bloodied fingermarks have been studied extensively using MALDI MS profiling and imaging previously [55–57] and an effective protocol has been developed:

1. Follow previous instructions depending on the type of fingermark required (ungroomed/groomed/natural/eccrine/contaminated).
2. Prick one finger of the non-deposition hand with a sterile lancet and apply pressure to force out a droplet of blood. Avoid transferring too much blood to the fingertip as this can cause smudging of the fingermark when deposited onto the surface, a clean pipette can be used to recover an appropriate amount of blood from the pricked finger.
3. Before the blood dries, rub the pricked finger and deposition finger together in order to distribute the blood across the fingertip.
4. Deposit a mark onto the desired surface.

3.3 Matrix Deposition

A range of MALDI matrices have been previously employed for the analysis of latent fingermarks, these include α-CHCA [18, 52, 54–66], dithranol [18, 53], curcumin [67], DHB [39, 51, 68], MBT [51], and 9-aminoacridine [51]. The selection of the MALDI matrix is dependent on the specific application required and the analytes of interest; therefore, optimization of this step may be necessary.

Before MALDI MS profiling or imaging can be conducted, the MALDI matrix must first be deposited onto the fingermark sample (*see* **Note 8**). There are now a range of methodologies available and a variety of sprayers that have emerged following the advent of MALDI MS imaging.

3.3.1 Spotting

1. Prepare the matrix solution of choice.
2. Pipette discrete spots of 0.5 μL of the MALDI matrix onto different areas of the fingermark.

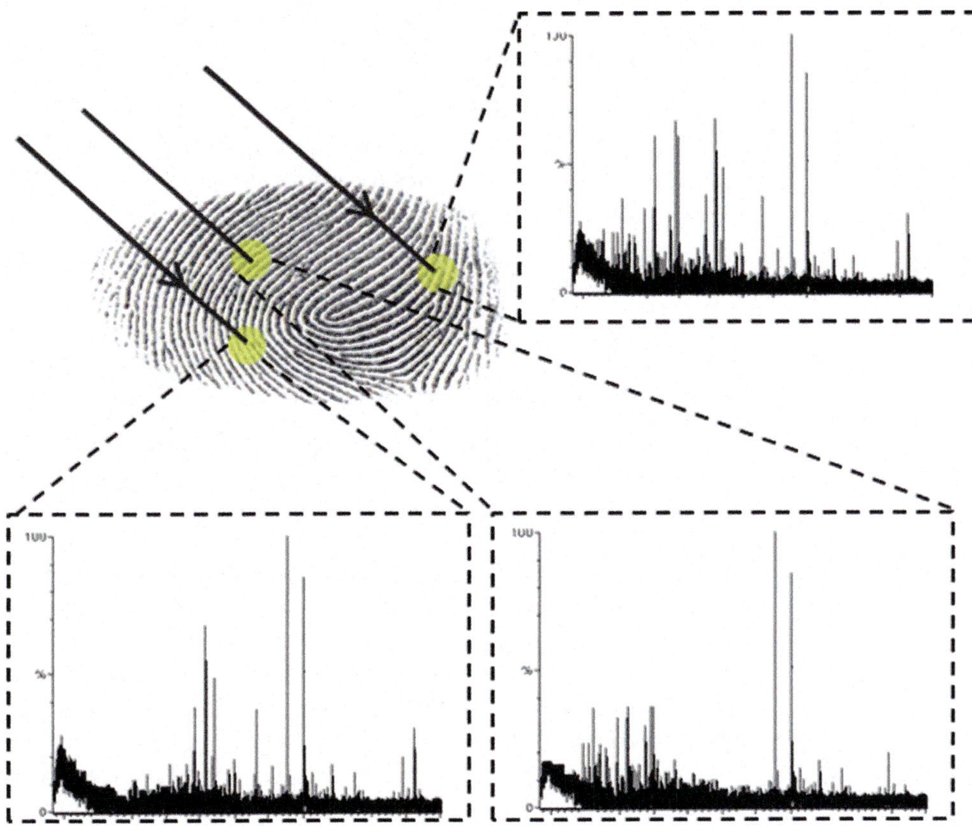

Fig. 1 Deposition of discrete matrix spots onto a fingermark sample and subsequent MALDI MS profiling showing how chemical differences can be observed between the selected areas

3. Provide adequate space between each of the matrix spots to ensure that they do not merge.
4. Allow the matrix spots to dry, this can take up to 30 min depending on the concentration of the matrix and the solvent employed.
5. Analyze the fingermark using MALDI MS profiling in the desired mass range—analyze each spot independently as the composition of specific fingermark components may not be homogenous across the entirety of the fingermark sample (Fig. 1).

3.3.2 Spraying

Suncollect Autosprayer

The "Suncollect" autosprayer (Sunchrom GmbH, Friedrichsdorf, Germany) employs a motorised nozzle linked to an inert gas line (such as N_2), producing a jet of matrix which is directed toward the sample. The flow rate of the matrix is dictated by a syringe driver, allowing for the reproducible deposition of matrices onto samples for MALDI MS Imaging applications. It has been used extensively for the preparation of fingermarks for MALDI MS analyses [18, 52, 54–58, 62–65].

1. Fill the syringe with solvent and flush the capillaries for at least 30 min at a flow rate of 2 μL/min.
2. Prepare the desired matrix solution.
3. Remove solvent from the syringe and fill with the matrix solution. Flush the capillaries for a further 15 min at a flow rate of 2 μL/min (*see* **Note 9**).
4. Set the coordinates of the sample (perimeter and height) using the instrumentation guidelines.
5. Turn on the gas flow and spray the sample (e.g., a total of five layers at a flow rate of 2 μL/min) using the "slow" raster setting.
6. Remove matrix from the syringe and fill with solvent. Flush the capillaries for at least 30 min at a flow rate of 2 μL/min (*see* **Note 10**).

HTX TM Sprayer

The HTX sprayer (HTX technologies, North Carolina, USA) employs a patented spray nozzle design that enables the deposition of the MALDI matrix at elevated temperatures (up to 140 °C). When combined with a high pressure system (often controlled by a HPLC pump) and optimized gas flow, rapid matrix deposition can be achieved with small crystal size facilitating high-resolution imaging. A novel 1 layer methodology was developed for the deposition of matrix onto fingermark samples, allowing for the spraying of a sample in under 1 min [66]. A more extensive outline of this methodology is as follows:

1. Prepare 400 mL of solvent in a 1 L wash bottle (this should be a similar composition of that used to dissolve the MALDI matrix (i.e., 70:30 ACN:Water)).
2. The wash bottle should be connected to the flow of the HPLC pump to allow continual purging of the sprayer.
3. Prepare 50 mL of the same solvent in a falcon tube (this will be used to flush the system prior to use).
4. Ensure that the system is set to the "load" position and flush with the prepared solvent (3 × 5 mL).
5. Change to the "spray" position and flush with the prepared solvent (3 × 5 mL).
6. Using the manufacturer's guidelines, create a method for the fingermark sample. An optimized methodology has been provided (Table 1).
7. Prepare the desired matrix solution.
8. Allow the nozzle to reach the specified temperature (e.g., 75 °C).
9. Load the MALDI matrix into the loop while in the "load" position (1 × 5 mL). Excess matrix will be redirected to waste.

Table 1
Optimized 1 min methodology for the deposition of matrix onto latent fingermarks using the HTX TM Sprayer [66]

N_2 pressure	Nozzle Temperature (°C)	Number of layers	Flow rate (mL/min)	Velocity (mm/min)	Track spacing (mm)
5	75	1	0.06	1300	1.5

10. Turn to the "spray" position and wait for matrix to be sprayed from the nozzle; this can be determined by holding a clean glass slide under the nozzle and waiting for matrix to be deposited.
11. Spray the sample.
12. Once spraying has completed, ensure that the system has been completely flushed with solvent by following **steps 4** and **5**.

Gravity Feed Airgun (Manual Sprayer)

The use of a gravity feed airgun is often necessary when depositing large amounts of matrix (volumes in excess of 5 mL), or when the chosen solvent/matrix is not compatible with an autosprayer system. Previously, manual spraying has been employed for the deposition of dithranol onto latent fingermark samples contaminated with condom lubricants [18, 53]. A more extensive outline for this methodology is provided below.

1. Clean the outside of the airgun thoroughly using approximately 5 mL ACN and absorbent tissue.
2. Load ACN into the airgun cup and flush thoroughly using at least 10 mL of solvent.
3. Line a mounting block (any surface that measures approximately 40 × 20 × 5 cm) with absorbent paper.
4. Secure the sample to the block with masking tape and stand it upright so that that the sample lies parallel to the sprayer nozzle.
5. Prepare 40 mL of the desired matrix solution.
6. Add the matrix solution to the airgun cup and seal the lid securely to avoid spillages.
7. At a distance of approximately 12 in from the sample, spray the matrix solution for 5 s, moving evenly across all areas of the sample during the time (*see* **Note 11**).
8. After 5 s, stop spraying and allow the sample to dry for approximately 20 s (this time may vary depending on the composition of the fingermark, the matrix solution employed, the pressure of the airflow, and the distance from the sample).
9. Continue this process until all 40 mL of the matrix solution has been deposited, or until the fingermark has been sufficiently coated with the MALDI matrix. This will require optimisation depending on the matrix, solvent and type of fingermarks being employed.

Fingermark Dusting (Dry-Wet Method)

The Dry-Wet method was first developed in 2011 [62] following a necessity to make the workflows more forensically applicable. In this process, the MALDI matrix (α-CHCA) is mechanically ground and sieved to produce crystals <10 μm in size. In the same manner in which latent fingermarks are developed using enhancement powders at crime scenes, the ground α-CHCA powder is dusted onto the fingermark samples (*see* **Note 12**) in order to provide visual development on non-porous and semi-porous surfaces. These developed marks can then be lifted using forensic lifting tape before being sprayed with solvent (to induce cocrystallization on the lifting tape) and analyzed by MALDI MS imaging.

1. Grind at least 1 g of α-CHCA using a mechanical ball mill or pestle and mortar.
2. Sieve the matrix using a 38 μm sieve.
3. Put a small amount of the dry matrix powder (approximately 100 mg) into the lid of a petri dish.
4. Collect the matrix onto the end of a zephyr fingerprint development brush, rotating the brush between the fingertips to ensure even distribution of the powder.
5. Using the same motion, rotate the brush between the fingertips while making light pressure with the substrate containing the latent fingermark.
6. After the powder has adhered to the fingermark ridges, remove excess powder using a Klenair air sprayer at a distance of approximately 30 cm from the fingermark.
7. Place forensic lifting tape directly onto the fingermark and apply a light pressure in order to adhere the matrix particles to the tape.
8. Peel the tape away from the substrate and transfer directly onto double-sided carbon attached to a MALDI target plate, ensuring that the adhesive side of the tape is facing upward.
9. Spray the fingermark using the Suncollect autosprayer, replacing the matrix solution with a solvent solution of 70:30 ACN:0.5% TFA (*see* **Note 13**).
10. The fingermark should be sprayed with a total of five layers at a flow rate of 5 μL/min using the "fast" raster setting.

4 Notes

1. The choice of matrix and solvent is heavily dependent upon the analytes of interest. The examples provided in this chapter should only be used as a general guide and further optimization may be necessary for different applications.

2. Aluminum sheets are recommended for fingermark deposition due to the ease of preparation, conductivity, ability to manipulate the material, flatness of the surface, and lack of porosity. However, this substrate is only provided as an example and alternatives can be employed depending on the application/outcomes required.
3. While preparing the fingermarks, the room temperature will greatly influence the extent of excretion from the fingertips; therefore, keep this consistent throughout all experiments.
4. Electronic scales can be used when depositing fingermarks to allow for consistent pressure to be achieved. A light/medium pressure should be used, corresponding to approximately 40–50 g (0.39–1.49 N). Excessive pressure will cause merging of the fingermark ridges and will affect the quality of the MALDI MS images generated.
5. When preparing groomed marks, refrain from using any cosmetics (such as hair products and moisturisers) on the day of deposition. These are likely to be transferred to the fingertips during the grooming procedure and could potentially be observed among the chemical species following MALDI MS analyses.
6. When employing the glass slide-transfer methodology for contaminated fingermarks, an appropriate solvent (such as MeOH) should be used to ensure rapid evaporation.
7. It is imperative to ensure that any exogenous substances used to contaminate fingermarks are safe to come into contact with the skin—if not then an artificial fingermark/press should be employed [65].
8. When using any matrix sprayers, ensure that the solvents employed are compatible by consulting the manufacturer's guidelines.
9. When flushing the matrix through the capillaries of the Suncollect autosprayer prior to spraying the sample, ensure that for the last 15 min, the flow rate is set the same as the rate that will be sprayed onto the fingermark. If the flow rate is set too high, back pressure may build up which can cause excessive matrix to be deposited onto the sample for the initial stages of the spraying sequence.
10. If multiple matrices are being deposited with the same sprayer, it may be necessary to flush the components for longer periods of time to avoid cross-contamination.
11. When employing the manual sprayer, it is necessary for the matrix to dry almost instantaneously after it is deposited onto the surface, thus reducing lateral diffusion of the fingermark ridges. However, the sample must also be wet enough to allow for extraction of analytes. Therefore, method optimization is necessary before preparing any samples.

12. Due to the generation of airborne dust particles, it is necessary to conduct the dusting of the fingermark samples within a fume hood and to wear a face mask at all times.
13. As the sample has been dusted with the dry MALDI matrix, spraying with solvent is necessary to allow for cocrystallization with the fingermark constituents.

Acknowledgments

The author of this chapter would like to thank the Head of the Fingermark Research Group (FRG), Dr Simona Francese, who has supervised every aspect of the fingermark research conducted at Sheffield Hallam University since the initial development of this application in 2008.

References

1. Champod C, Lennard CJ, Margot P, Stoilovic M (2004) Fingerprints and other ridge skin impressions. CRC Press LLC, London
2. Sears V, Bleay S, Bandey H, Bowman V (2012) A methodology for finger mark research. Sci Justice 52:145–160
3. Girod A, Weyermann C (2014) Lipid composition of fingermark residue and donor classification using GC/MS. Forensic Sci Int 238:68–82
4. Leggett R, Lee-Smith E, Jickells S, Russell D (2007) "Intelligent" fingerprinting: simultaneous identification of drug metabolites and individuals by using antibody-functionalized nanoparticles. Angew Chem Int Ed 46:4100–4103
5. Hazarika P, Jickells S, Wolff K, Russell D (2008) Imaging of latent fingerprints through the detection of drugs and metabolites. Angew Chem 120:10321–10324
6. Drapel V, Becue A, Champod C, Margot P (2009) Identification of promising antigenic components in latent fingermark residues. Forensic Sci Int 184:47–53
7. Spindler X, Hofstetter O, McDonagh A, Roux C, Lennard C (2011) Enhancement of latent fingermarks on non-porous surfaces using anti-L-amino acid antibodies conjugated to gold nanoparticles. Chem Commun 47:5602–5604
8. Van Dam A, Aalders M, Van Leeuwen T, Lambrechts S (2013) The compatibility of fingerprint visualization techniques with immunolabeling. J Forensic Sci 58(4):999–1002
9. Van Dam A, Aalders M, De Puit M, Gorré S, Irmak D, Van Leeuwen T, Lambrechts S (2014) Immunolabeling and the compatibility with a variety of fingermark development techniques. Sci Justice 54:356–362
10. Van der Heide S, Calavia P, Hardwick S, Hudson S, Wolff K, Russell D (2015) A competitive enzyme immunoassay for the quantitative detection of cocaine from banknotes and latent fingermarks. Forensic Sci Int 250:1–7
11. Lam R, Hofstetter O, Lennard C, Roux C, Spindler X (2016) Evaluation of multi-target immunogenic reagents for the detection of latent and body fluid-contaminated fingermarks. Forensic Sci Int 264:168–175
12. Day J, Edwards H, Dobrowski S, Voice A (2004a) The detection of drugs of abuse in fingerprints using Raman spectroscopy I: latent fingerprints. Spectrochim Acta A Mol Biomol Spectrosc 60:563–568
13. Day J, Edwards H, Dobrowski S, Voice A (2004b) The detection of drugs of abuse in fingerprints using Raman spectroscopy II: cyanoacrylate-fumed fingerprints. Spectrochim Acta A Mol Biomol Spectrosc 60:1725–1730
14. Tahtouh M, Kalman J, Roux C, Lennard C, Reedy B (2005) The detection and enhancement of latent fingermarks using infrared chemical imaging. J Forensic Sci 50(1):1–9
15. West M, Went M (2008) The spectroscopic detection of exogenous material in fingerprints after development with powders and recovery with adhesive lifters. Forensic Sci Int 174:1–5
16. West M, Went M (2009) The spectroscopic detection of drugs of abuse in fingerprints after development with powders and recovery with

adhesive lifters. Spectrochim Acta A Mol Biomol Spectrosc 7:1984–1988

17. Widjaja E (2009) Latent fingerprints analysis using tape-lift, Raman microscopy, and multivariate data analysis methods. Analyst 134:769–775
18. Bradshaw R, Wolstenholme R, Ferguson L, Mader K, Sammon C, Blackledge RD, Clench MR, Francese S (2013a) Spectroscopic Imaging based approach for condom identification in condom contaminated fingermarks. Analyst 138(9):2546–2557
19. Banas A, Banas K, Breese B, Loke J, Lim S (2014) Spectroscopic detection of exogenous materials in latent fingerprints treated with powders and lifted off with adhesive tapes. Anal Bioanal Chem 17:4178–4181
20. Girod A, Xiao L, Reedy B, Roux C, Weyermann C (2015) Fingermark initial composition and aging using Fourier transform infrared microscopy (m-FTIR). Forensic Sci Int 254:185–196
21. Archer N, Charles Y, Elliott J, Jickells S (2005) Changes in the lipid composition of latent fingerprint residue with time after deposition on a surface. Forensic Sci Int 154:224–239
22. Croxton R, Baron M, Butler D, Kent T, Sears V (2010) Variation in amino acid and lipid composition of latent fingerprints. Forensic Sci Int 199:93–102
23. Weyermann C, Roux C, Champod C (2011) Initial results on the composition of fingerprints and its evolution as a function of time by GC/MS analysis. J Forensic Sci 56:102–108
24. Michalski S, Shaler R, Dorman F (2013) The evaluation of fatty acid ratios in latent fingermarks by gas chromatography/mass spectrometry (GC/MS) analysis. J Forensic Sci 58:215–220
25. Mink T, Voorhaar A, Stoel R, de Puit M (2013) Determination of efficacy of fingermark enhancement reagents; the use of propyl chloroformate for the derivatization of fingerprint amino acids extracted from paper. Sci Justice 53:301–308
26. Frick A, Chidlow G, Lewis S, van Bronswijk W (2015) Investigations into the initial composition of latent fingermark lipids by gas chromatography–mass spectrometry. Forensic Sci Int 254:133–147
27. Girod A, Spyratou A, Holmes D, Weyermann C (2016) Aging of target lipid parameters in fingermark residue using GC/MS: effects of influence factors and perspectives for dating purposes. Sci Justice 56:165–180
28. Zhang T, Chen X, Yang R, Xu Y (2015) Detection of methamphetamine and its main metabolite in fingermarks by liquid chromatography–mass spectrometry. Forensic Sci Int 248:10–14
29. Ifa D, Manicke N, Dill A, Cooks G (2008) Latent fingerprint chemical imaging by mass spectrometry. Science 321:805
30. Mirabelli M, Chramow A, Cabral E, Ifa D (2011) Analysis of sexual assault evidence by desorption electrospray ionization mass spectrometry. J Mass Spectrom 48:774–778
31. Comi T, Ryu S, Perry R (2016) Synchronized desorption electrospray ionization mass spectrometry imaging. Anal Chem 88:1169–1175
32. Rowell F, Hudson K, Seviour J (2009) Detection of drugs and their metabolites in dusted latent fingermarks by mass spectrometry. Analyst 134:701–707
33. Benton M, Rowell F, Sundar L, Jan M (2010a) Direct detection of nicotine and cotinine in dusted latent fingermarks of smokers by using hydrophobic silica particles and MS. Surf Interface Anal 42:378–385
34. Benton M, Chu M, Gu F, Rowell F, Ma J (2010b) Environmental nicotine contamination in latent fingermarks from smoker contacts and passive smoking. Forensic Sci Int 200:28–34
35. Lim A, Ma Z, Ma J, Rowell F (2011) Separation of fingerprint constituents using magnetic silica nanoparticles and direct on-particle SALDI-TOF-mass spectrometry. J Chromatogr B 879:2244–2250
36. Lim A, Seviour J (2012) Doped silica nanoparticles for the detection of pharmaceutical Terbinafine in latent fingerprints by mass spectrometry. Anal Methods 4:1983–1198
37. Rowell F, Seviour J, Lim A, Elumbaring-Salazar C, Loke J, Ma J (2012) Detection of nitro-organic and peroxide explosives in latent fingermarks by DART- and SALDI-TOF-mass spectrometry. Forensic Sci Int 221:84–91
38. Nizioł J, Ruman T (2013) Surface-transfer mass spectrometry imaging on a monoisotopic silver nanoparticle enhanced target. Anal Chem 85:12070–12076
39. Sundar L, Rowell F (2014) Detection of drugs in lifted cyanoacrylate developed latent fingermarks using two laser desorption/ionisation mass spectrometric methods. Analyst 139:633–642
40. Tang H, Lu W, Che C, Ng K (2010) Gold nanoparticles and imaging mass spectrometry: double imaging of latent fingerprints. Lett Anal Chem 82:1589–1593
41. Szynkowska M, Czerskia K, Gramsa J, Paryjczaka T, Parczewskib A (2007) Preliminary studies using imaging mass spectrometry TOF-SIMS in detection and analysis of fingerprints. Imag Sci J 55(3):180–187

42. Szynkowska M, Czerski K, Rogowski J, Paryjczak T, Parczewski A (2009) ToF-SIMS application in the visualization and analysis of fingerprints after contact with amphetamine drugs. Forensic Sci Int 184:24
43. Szynkowska M, Czerski K, Rogowski J, Paryjczaka T, Parczewski A (2010) Detection of exogenous contaminants of fingerprints using ToF-SIMS. Surf Interface Anal 42:393
44. Bailey M, Jones B, Hinder S, Watts J, Bleay S, Webb R (2010) Depth profiling of fingerprint and ink signals by SIMS and MeV SIMS. Nucl Instrum Methods Phys Res, Sect B 268:1929–1932
45. Bright N, Willson T, Driscoll D, Reddy S, Webb R, Bleay S, Ward N, Kirkby K, Bailey M (2013) Chemical changes exhibited by latent fingerprints after exposure to vacuum conditions. Forensic Sci Int 230:81–86
46. Attard-Montalto N, Ojeda J, Jones B (2013) Determining the order of deposition of natural latent fingerprints and laser printed ink using chemical mapping with secondary ion mass spectrometry. Sci Justice 53:2–7
47. Bailey M, Ismail M, Bleay S, Bright N, Elad N, Cohen Y, Geller B, Everson D, Costa C, Webb R, Watts J, de Puit M (2013) Enhanced imaging of developed fingerprints using mass spectrometry imaging. Analyst 138: 6246–6250
48. Musah R, Cody R, Dane A, Vuong A, Shepard J (2012) Direct analysis in real time mass spectrometry for analysis of sexual assault evidence. Rapid Commun Mass Spectrom 26:1039–1046
49. Emerson B, Gidden J, Lay J Jr, Durham B (2011) Laser desorption/ionization time-of-flight mass spectrometry of triacylglycerols and other components in fingermark samples. J Forensic Sci 56(2):381–389
50. Lauzon N, Dufresne M, Chauhan V, Chaurand P (2015) Development of laser desorption imaging mass spectrometry methods to investigate the molecular composition of latent fingermarks. J Am Soc Mass Spectrom 26(6):878–886
51. Francese S (2015) Techniques for fingermark analysis using MALDI MS - a practical overview. In: Cramer R (ed) Advances in MALDI and laser induced soft ionisation mass spectrometry. Springer, New York, NY. isbn:978-3-319-04818-5
52. Wolstenholme R, Bradshaw R, Clench M, Francese S (2009) Study of latent fingermarks by matrix-assisted laser desorption/ionisation mass spectrometry imaging of endogenous lipids. Rapid Commun Mass Spectrom 23:3031–3039
53. Bradshaw R, Wolstenholme R, Blackledge R, Clench MR, Ferguson L, Francese S (2011) A novel matrix-assisted laser desorption/ionisation mass spectrometry imaging based methodology for the identification of sexual assault suspects. Rapid Commun Mass Spectrom 25:415–422
54. Bradshaw R, Rao W, Wolstenholme R, Clench MR, Bleay S, Francese S (2012) Separation of overlapping fingermarks by matrix assisted laser desorption ionisation mass spectrometry imaging. Forensic Sci Int 222(1-3):318–326
55. Bradshaw R, Bleay S, Clench MR, Francese S (2014) Direct detection of blood in fingermarks by MALDI MS profiling and Imaging. Sci Justice 54:110–117
56. Patel E, Cicatiello P, Deininger L, Clench M, Marino G, Giardina P, Langenburg G, West A, Marshall P, Searse V, Francese S (2016) A proteomic approach for the rapid, multi-informative and reliable identification of blood. Analyst 141:191–198
57. Deininger L, Patel E, Clench M, Sears V, Sammon C, Francese S (2016) Proteomics goes forensic: detection and mapping of blood signatures in fingermarks. Proteomics 16(11-12): 1707–1717
58. Groeneveld G, De Puit M, Bleay S, Bradshaw R, Francese S (2015) Detection and mapping of illicit drugs and their metabolites in fingermarks by MALDI MS and compatibility with forensic techniques. Sci Rep 5:11716
59. Bailey M, Bradshaw R, Francese S, Salter T, Costa C, Ismail M, Webb R, Bosman I, Wolff K, de Puit M (2015) Rapid detection of cocaine, benzoylecgonine and methylecgonine in fingerprints using surface mass spectrometry. Analyst 140(18):6254–6259
60. Kaplan-Sandquist K, LeBeau M, Miller M (2014) Chemical analysis of pharmaceuticals and explosives in fingermarks using matrix-assisted laser desorption ionization/time-of-flight mass spectrometry. Forensic Sci Int 235:68–77
61. Ferguson L, Wulfert F, Wolstenholme R, Fonville J, Clench MR, Carolan V, Francese S (2012) Direct detection of peptides and small proteins in fingermarks and determination of sex by MALDI mass spectrometry profiling. Analyst 137:4686–4692
62. Ferguson L, Bradshaw R, Wolstenholme R, Clench M, Francese S (2011) Two-step matrix application for the enhancement and imaging of latent fingermarks. Anal Chem 83(14): 5585–5591
63. Ferguson L, Creasey S, Wolstenholme R, Clench M, Francese S (2012) Efficiency of the dry-wet method for the MALDI-MSI analysis of latent fingermarks. J Mass Spectrom 48:677–684

64. Bradshaw R, Bleay S, Wolstenholme R, Clench MR, Francese S (2013b) Towards the integration of matrix assisted laser desorption ionisation mass spectrometry imaging into the current fingermark examination workflow. Forensic Sci Int 232:111–124
65. Reed H, Stanton A, Wheat J, Kelley J, Davis L, Rao A, Smith A, Owen D, Francese S (2016) The Reed-Stanton press rig for the generation of reproducible fingermarks: towards a standardised methodology for fingermark research. Sci Justice 56:9–17
66. Bradshaw R, Creissen A, Francese S (2013c), A rapid methodology for the analysis of latent fingermarks by MALDI MS imaging. Technical Note #32, HTX Technologies
67. Francese S, Bradshaw R, Flinders B, Mitchell C, Bleay S, Cicero L, Clench R (2013) Curcumin: a multipurpose matrix for MALDI mass spectrometry imaging applications. Anal Chem 85(10):5240–5248
68. Yagnik G, Kortea A, Lee Y (2013) Multiplex mass spectrometry imaging for latent fingerprints. J Mass Spectrom 48:100–104

Chapter 14

ToF-SIMS Parallel Imaging MS/MS of Lipid Species in Thin Tissue Sections

Anne Lisa Bruinen, Gregory L. Fisher, and Ron M.A. Heeren

Abstract

Unambiguous identification of detected species is essential in complex biomedical samples. To date, there are not many mass spectrometry imaging techniques that can provide both high spatial resolution and identification capabilities. A new and patented imaging tandem mass spectrometer, exploiting the unique characteristics of the *nanoTOF* II (Physical Electronics, USA) TOF-SIMS TRIFT instrument, was developed to address this.

Tandem mass spectrometry is based on the selection of precursor ions from the full secondary ion spectrum (MS^1), followed by energetic activation and fragmentation, and collection of the fragment ions to obtain a tandem MS spectrum (MS^2). The PHI *NanoTOF II* mass spectrometer is equipped with a high-energy collision induced dissociation (CID) fragmentation cell as well as a second time-of-flight analyzer developed for simultaneous ToF-SIMS and tandem MS imaging experiments.

We describe here the results of a ToF-SIMS imaging experiment on a thin tissue section of an infected zebrafish as a model organism for tuberculosis. The focus is on the obtained ion distribution plot of a fatty acid as well as its identification by tandem mass spectrometry.

Key words SIMS, Tandem MS, Imaging mass spectrometry, Lipids, Fatty acids

1 Introduction

Time-of-flight secondary ion mass spectrometry (ToF-SIMS) analysis depends on a high energy focused ion beam that functions as a primary ion source ejecting analytes only from the top atomic monolayers of a sample surface, referred to as secondary ions. The secondary ions are separated in a time-of-flight mass analyzer based on their mass-to-charge ratio (m/z) before they hit the detector [1–3].

Traditionally, ToF-SIMS has been applied in the domain of surface physics and semiconductor industry, but over the last decade it has redirected toward biomedical applications as well. Especially in mass spectrometry imaging-based biological tissue analysis, one has to contend with a high level of complexity concerning the molecules that are found on the sample surface. In many cases, it is predictable and generally accepted which

Laura M. Cole (ed.), *Imaging Mass Spectrometry: Methods and Protocols*, Methods in Molecular Biology, vol. 1618, DOI 10.1007/978-1-4939-7051-3_14, © Springer Science+Business Media LLC 2017

compounds are detected in specific samples. Although the mass resolution and mass accuracy in ToF analyzers have improved over the years, particularly for higher masses this is not sufficient for unambiguous peak identification.

Tandem mass spectrometry is considered the benchmark technique to identify and elucidate structures of unknown molecular species and distinguish isomers of which there are multiplicities in biological samples. Controlled fragmentation of ions after they are formed in the source of a mass spectrometer results in a unique fragment pattern for every compound and is therefore an extremely useful tool for identification and elucidation of molecules.

Matrix-assisted laser desorption/ionization (MALDI) is often the ionization technique of choice for tandem MS imaging in biological tissues, but has several inherent limitations compared to ToF-SIMS imaging. First, in contrast to ToF-SIMS, MALDI needs a homogeneously applied chemical matrix to absorb the laser energy, to extract and ionize compounds from the sample. This makes the sample preparation time consuming and results in the perturbation of the intrinsic molecular chemistry and loss of lateral resolution. Furthermore, photon ionization of the analytes in a MALDI ion source is very destructive and usually consumes the sample, whereas in ToF-SIMS only sub nanometer thick layers are removed from the sample surface. In addition, the typical spatial resolution of MALDI imaging is above 30 μm, ToF-SIMS imaging can reach submicron spatial resolution. Lastly, current instrument designs do not allow simultaneous collection of MS^1 and MS^2 imaging data sets.

Recently, a novel and patented PHI *NanoTOF II* TRIFT mass spectrometer was developed, equipped with a second time-of-flight analyzer designed for tandem MS imaging [4–6]. The ion optics of the first mass analyser allow selecting a precursor from the stream of generated ions to be analyzed in the tandem time-of-flight mass analyzer simultaneously. This makes the combination of surface screening with general MS imaging with targeted identification with tandem MS imaging possible (Fig. 1).

In this chapter, we show an example of a tandem TOF-SIMS imaging experiment performed on a whole-body section prepared from a diseased zebrafish. The fish was part of a study on tuberculosis and therefore infected with *Mycobacterium marinum* as a model for human tuberculosis infection. Without further preparation, the samples were analyzed in negative ion mode using a PHI *nanoTOF* II TOF-TOF mass spectrometer equipped with a liquid metal ion gun (Bi_3^+), a gas collision cell, and a second ToF detector. ToF-SIMS tandem imaging MS was used to localize and identify lipid species and fatty acids as well as the host membrane phospholipids that are believed to play an important role in the bacterial inflammation cascade.

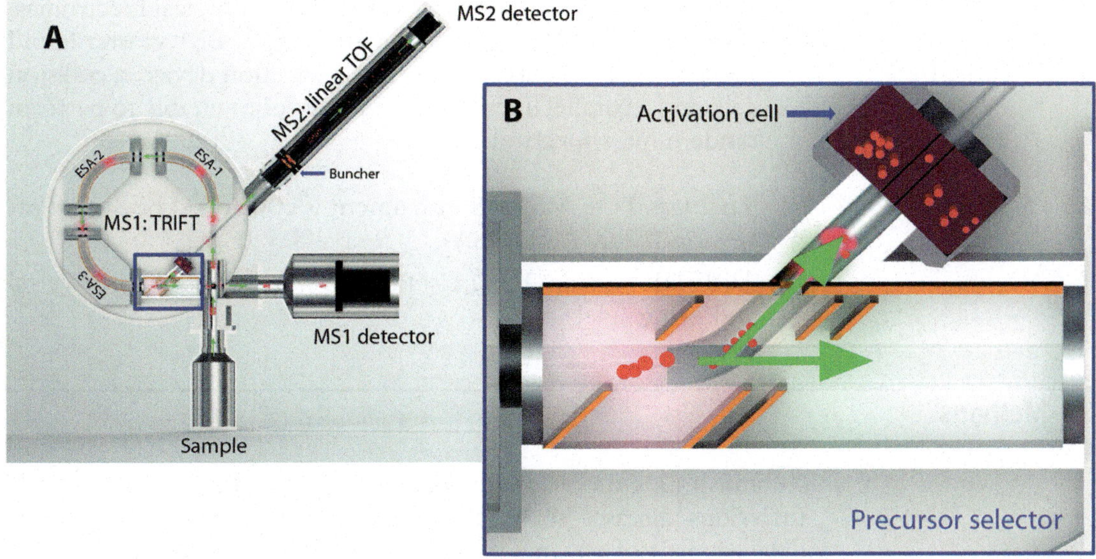

Fig. 1 (**a**) Schematic illustration of the parallel imaging MS/MS spectrometer with the MS1 (TRIFT) and the MS2 (linear TOF) analysers as well as their detectors annotated. The arrows indicate the nominal trajectories of the secondary ions (in red). The precursor selection device including the activation cell is framed in (**a**) and enlarged in (**b**)

2 Materials

2.1 Chemicals

1. Dietrich's fixative (30% ethanol, 10% formalin, 2% glacial acetic acid in deionized water).
2. Mixture of 5% carboxyl methyl cellulose (~90,000 Da polymer length) and 10% porcine gelatine, kept liquid at 37 °C.
3. Dry ice.
4. Ethanol.
5. Indium tin oxide (ITO)-coated glass slides.
6. Starfrost glass slides.
7. Acetone (at −20 °C).
8. Eosin.
9. Hematoxylin.
10. MilliQ water.
11. 95% ethanol.
12. Pure ethanol (96%).
13. Glycerol.
14. Cover slip.

2.2 Instrumentation

1. Cryotome at −35 °C.
2. Mirax slide scanner (Carl Zeiss).

3. PHI *nano*TOF II TOF-SIMS instrument (Physical Electronics, Minnesota, U.S.A.) equipped with a 30 kV Bi_n^{q+} cluster liquid metal ion gun (LMIG), a precursor selection device, a collision cell, and a parallel linear time-of-flight tube suitable to perform tandem mass spectrometry.

2.3 Software

1. The *nano*TOF MS/MS instrument is controlled by SmartSoft version 2.1.0 (PHI, USA).
2. TOF-DR version 1.1.0.1. (PHI, USA) was used for spectral and image analysis.

3 Methods

Zebrafish (*Danio rerio*) are considered a suitable model to study infectious diseases such as tuberculosis (TB) and to aid in the development of treatments [7]. *Mycobacterium marinum* bacteria cause tuberculosis in fish, and has a very similar disease process compared to *Mycobacterium tuberculosis*, which is responsible for tuberculosis infections in humans.

Foamy macrophages are loaded with lipids and use phagocytosis with the aim to inactivate the bacteria, but TB bacteria use these cells to replicate. That causes more macrophages to attack the infected ones and by that the formation of granulomas [8, 9]. The outcome of tuberculosis is different for every individual, bacteria strand, and possible treatment.

In this study, we seek to develop a way to visualize and reveal the content of tuberculosis granulomas. The mapping of molecular changes captured with mass spectrometry imaging gives more insight into the immune response of the host (zebrafish) toward the pathogen (*M. marinum*). Because the *nano*TOF system facilitates high lateral resolution molecular imaging, lipids within the granuloma structures can be visualized. Furthermore, because of the nondestructive nature of SIMS imaging, tandem MS imaging of the same sample is possible for multiple precursor ions repeatedly.

3.1 Sample Preparation

An adult zebrafish (*D. rerio*) was infected by *M. marinum* bacteria (*mptC* mutant strain [10]) and killed as part of an infection study [11].

1. Fix the fish chemically in Dietrich's fixative for >48 h at 4 °C. Fixation induces decontamination of all bacteria, which enables a safer working environment.
2. Take the fish out of the fixative and rinse it with MilliQ water a couple of times to get rid of excess fixative.
3. Embed the specimen in a mixture of 5% carboxyl methyl cellulose and 10% gelatin in a mold [12] (*see* **Note 1**).

4. Freeze the mold including the embedding material with the specimen in a 1:1 dry ice/ethanol bath. It takes about 15 min for the block to freeze, indicated by the ice turning white.
5. Leave the block in a freezer (−20 °C/−80 °C) overnight to completely solidify.
6. Prior to cutting, place the block containing the specimen in the cryotome at −35 °C for 1.5 h to adjust to the cutting temperature.
7. Fix the block to the cryotome stage with a droplet of warm 5% carboxyl methyl cellulose and 10% gelatine.
8. Cut 20 μm longitudinal tissue sections and thaw mount them on the conductive side of indium tin oxide-coated glass slides (*see* **Notes 2** and **3**).
9. Choose a tissue section collected on a conductive ITO slide showing multiple organs as well as severe infection areas for TOF-SIMS analysis. Prepare an adjacent section and thaw mount it on a Starfrost slide to have it stick better and make it suitable for histological staining.
10. Place the Starfrost slide in an acetone solution at −20 °C to fix for 5 min.
11. Perform a hematoxylin and eosin (H&E) staining on the adjacent section on a Starfrost slide to show contrast as follows:
    (a) Rinse the slide in MilliQ water.
    (b) Stain in hematoxylin solution for 30 s.
    (c) Rinse in MilliQ water for 30 s.
    (d) Wash in 95% ethanol for 30 s.
    (e) Stain in eosin for 30 s.
    (f) Rinse in pure ethanol (96%) for 2 min (two times).
    (g) Air dry the tissue.
    (h) Mount a cover slip with glycerol.
12. Scan the H&E stained slide with a Mirax slide scanner (Carl Zeiss). The image obtained from the described study is shown in Fig. 2a with annotations.

3.2 Tandem MS Imaging

The fundamentals of the modified PHI *NanoTOF II* tandem MS mass spectrometer used for this experiment are based upon the Triple Ion Focussing Time-of-flight (TRIFT) mass analyzer introduced by Schueler et al. in 1990 [13, 14]. It concerns a three-sector mass spectrometer that separates ions based on their mass-to-charge ratio (m/z). A 30 kV Bi_n^{q+} cluster liquid metal ion gun (LMIG) produces secondary ions by striking the sample surface and sputters them off.

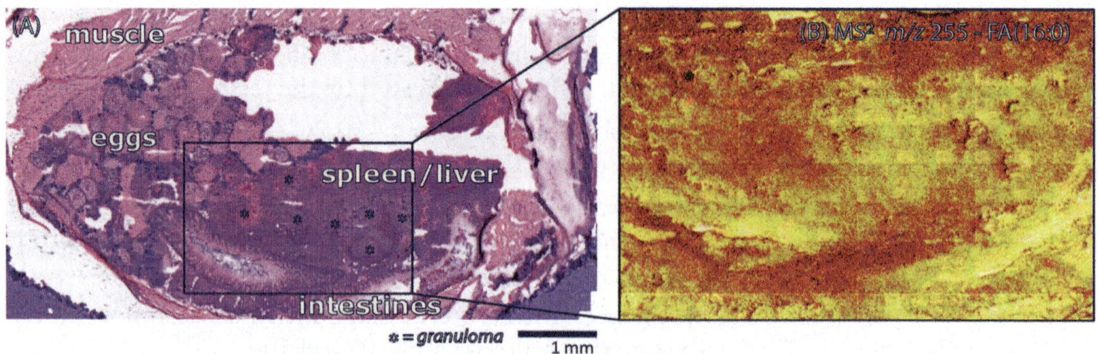

Fig. 2 (**a**) Histology of the intestine area of a zebrafish infected by Mycobacterium marinum. The different organs are annotated by their names, and several granulomatous structures appearing in the spleen and liver area are annotated by an asterisk. The box indicates the measurement region containing several granulomata from which an ion distribution plot is generated. (**b**) Logarithmic scale MS^2 mosaic map of the total ion count (TIC) of an infected zebrafish spleen area, containing the selected precursor (m /z 255) and its fragments. The dimensions are 4.8 mm × 3.2 mm

To achieve tandem MS imaging, a PHI *nanoTOF* II (Physical Electronics, USA) TOF-SIMS TRIFT analyzer was modified. The original blanker located after the existing mass separation path of the secondary ions through three electrostatic analyzers (ESAs) was replaced by a precursor selection device. This is an assembly of three electrode plates of which the two outer ones are held at ground voltage, while the middle electrode can be pulsed electrically. In the middle of every plate a small aperture is located, through which the secondary ions travel after mass-to-charge separation. The timing of the pulse of the middle electrode is crucial. To pick ions with a particular mass from the secondary ion stream, the second electrode generates a pulse exactly when the ions are located between the first outer and the middle electrode. Only then precursor ions can be deflected parallel to the electrode plates, 45° from the secondary ion stream, into the fragmentation cell and MS2 mass analyser, while the rest of the secondary ions travel through toward the MS1 detector. The precursor selection device is able to deflect ions of a specific *m/z* from 0 up to 100% of the pulses passing by, but also functions as a post ESA blanker and deflect a larger mass range.

To provide fragmentation, the deflected precursor ions are guided through a collision cell filled with argon. This collision-induced dissociation (CID) produces fragment ions, but also energy spread among them. To correct for this, a buncher constructed of two parallel grids is placed right after the collision cell. The grids are kept at ground potential, until a pulse of fragment ions is present in the space between the grids. A voltage is then applied to the first grid to accelerate the ions toward the direction of the second grid. Mass-to-charge separation of the accelerated fragment ions is accomplished in a field free region followed by a dual multichannel plate (DMCP)-based detector, referred to as the MS2 detector. This setup is similar to the MS1 detector.

1. Optimize the 30 keV bismuth LMIG on the PHI *nanoTOF* II for molecular imaging. Select Bi_3^+ clusters and operate the gun in bunched mode for best spectral quality. A typical tuning copper grid (mesh 25 μm) on an aluminum substrate can be used to optimize the beam current as well as the image quality by fine-tuning the focussing lenses in the gun.
2. Make sure the sample bias is at the optimal voltage and the Z-height is correct as well for the specific sample by starting an acquisition. Both parameters are optimal when the secondary ion yield is highest and hit the middle of the detector. The ion count is easily visible in the SmartSoft software that is linked to the CCD camera in the detector assembly.
3. While acquiring from the sample of interest, it is very important to calibrate the MS^1 spectrum before moving toward tandem MS experiments. This can be done online in the TOF-DR software with the use of known peaks (*see* **Note 4**).
4. To perform tandem MS, open the gas leak valve to the CID activation cell to fill it with argon gas. The desired precursor peak can be selected in the SmartSoft software, thereafter the approximate values concerning the selection timing and post ESA slope are set. However, in most cases, these settings need to be fine tuned for each different *m/z* value. For the described experiment *m/z* 255.230 was selected, the post ESA slope was set at 2.0487, the picker voltage at 2780, the picker width at 0.702, and the buncher at 1300 V.
5. To be able to generate molecular images, the primary ion beam is rastered over a 400 μm × 400 μm area (tile), containing 256 × 256 pixels. The whole measured area is 4.8 mm × 3.2 mm in size, containing 12 × 8 of these tiles, exposed for five frames each.
6. The acquisition is saved as a single .raw file, containing both MS^1 and MS^2 data acquired simultaneously. In the TOF-DR software it is possible to playback a measurement and select single mass channels to generate two-dimensional ion distribution plots.

The precursor selection device was set to deflect only ions with a mass-to-charge ratio of 255 to the MS2 detector. The image of the total MS^2 ion count is shown in Fig. 2b, representing the spatial distribution of all ions with *m/z* 255.

The molecular fragments originating from the *m/z* 255 precursor are captured in the tandem MS spectrum shown in Fig. 3. Since the precursor peak is the most abundant peak in the spectrum, the mass region representing the most important fragments

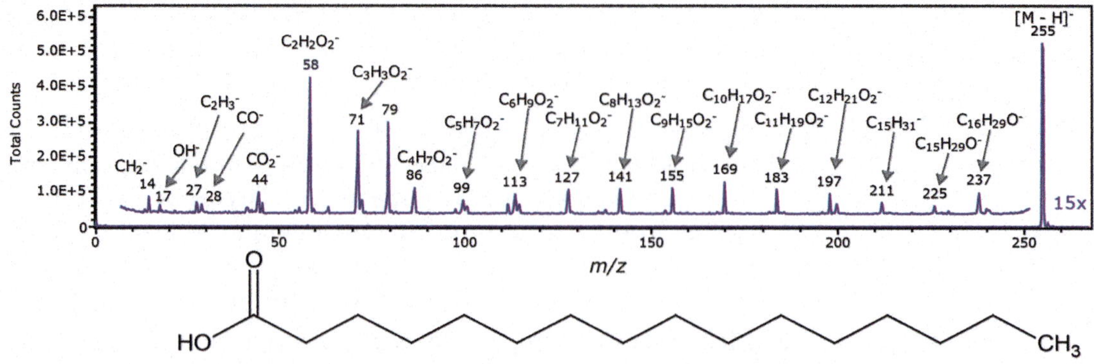

Fig. 3 CID product ion spectrum of the m /z 255 precursor produced from the surface of a thin tissue section sampled from a M. Marinum infected adult zebrafish. The precursor was identified as palmitic acid (FA (16:0)) of which the structural formula is shown below the tandem MS spectrum

is zoomed in. The typical CH_2 loss sequence shows that this spectrum undeniably originates from the long carbon chain of a fatty acid, namely palmitic acid.

4 Notes

1. Embedding of a zebrafish is carried out nicely in an ice cube tray. Make sure there is enough space around the specimen for embedding material to prevent defrosting when taken out of the freezer. To keep the fish straight, it is recommended to pour in a layer of embedding mixture first and let it solidify at room temperature. Then place the fish on its side on the top of the material and cover it with more embedding mixture.
2. Longitudinal tissue sections give a more representative overview of different organs, but the cutting of the block is easier in a vertical manner (starting at the long side of the fish) to prevent ruptures inside the tissue section. This is due to the fact that fish contain cavities filled with air, among others a swim bladder. Although 20 μm sections seem to be thick for imaging mass spectrometry, it is the optimal thickness to keep the morphology of the fish intact. Cutting a whole organism is challenging since one has to deal with a variety of tissue densities.
3. When cutting multiple consecutive tissue sections, it is recommended to leave the block to cool for about 10 min frequently. The surface temperature rises by touching it with the knife during cutting, which has a negative influence on the section quality.
4. When analyzing organic molecules from the sample surface, it is advisable to calibrate the spectrum with known organic

peaks. For biological tissue sections typical peaks are *m/z* 15.0235 (CH_3), 27.0235 (C_2H_3), 41.0391 (C_3H_5), 55.0547 (C_4H_7), 69.0704 (C_5H_9), and 86.0969 ($C_5H_{12}N$), 184.0737 ($C_5H_{15}O_4NP$). In negative mode the following *m/z* are very suitable: 15.9949 (O), 26.0031 (CN), 41.9980 (CNO), and 78.9585 (PO_3).

References

1. Pachuta SJ, Cooks RG (1987) Mechanisms in molecular SIMS. Chem Rev 87(3): 647–669
2. Castaing R, Slodzian G (1962) Microanalyse par emission ionique secondaire. J Microsc 1:31–38
3. Werner HW (1975) The use of secondary ion mass spectrometry in surface analysis. Surf Sci 47(1):301–323
4. Larson, PE, et al., (2013) Method and apparatus to provide parallel acquisition of mass spectrometry/mass spectrometry data. 2013: United States.
5. Fisher GL et al (2016) Parallel imaging MS/MS TOF-SIMS instrument. J Vac Sci Technol B 34(3):03H126
6. Fisher GL et al (2016) A new method and mass spectrometer design for TOF-SIMS parallel imaging MS/MS. Anal Chem 88(12): 6433–6440
7. van der Sar AM et al (2004) A star with stripes: zebrafish as an infection model. Trends Microbiol 12(10):451–457
8. Toossi Z (2000) The inflammatory response in mycobacterium tuberculosis infection. Arch Immunol Ther Exp (Warsz) 48:513–519
9. Adams DO (1976) The granulomatous inflammatory response. A review. Am J Pathol 84: 164–192
10. Stoop EJM et al (2013) Mannan core branching of lipo(arabino)mannan is required for mycobacterial virulence in the context of innate immunity. Cell Microbiol 15(12):2093–2108
11. van der Sar AM et al (2004) Mycobacterium marinum strains can be divided into two distinct types based on genetic diversity and virulence. Infect Immun 72(11):6306–6312
12. Nelson K et al (2013) Optimization of whole-body zebrafish sectioning methods for mass spectrometry imaging. J Biomol Tech 24(3): 119–127
13. Schueler B, Sander P, Reed DA (1990) A time-of-flight secondary ion microscope. Vacuum 41(7–9):1661–1664
14. Schueler BW (1992) Microscope imaging by time-of-flight secondary ion mass spectrometry. Microsc Microanal Microstruct 3(2-3):119–139

Chapter 15

Rodent Whole-Body Sectioning and MALDI Mass Spectrometry Imaging

Paul J. Trim

Abstract

Here, we describe a method for obtaining whole-body MALDI imaging data. MALDI imaging provides chemical compound-specific information not attainable with conventional histology techniques. The specificity of mass spectrometry with the addition of spatial information makes this a very powerful technique, especially for the analysis of endogenous and exogenous small molecules. This chapter will provide the reader with a comprehensive description of the techniques involved in obtaining high-quality MALDI mass spectrometry imaging (MSI) data from large tissue sections.

Key words MALDI imaging mass spectrometry, Whole-body, QWBA

1 Introduction

This chapter provides a practical guide to sample preparation and subsequent analysis of small molecule drugs, metabolites, or endogenous compounds for whole-body tissue sections using matrix-assisted laser desorption/ionisation mass spectrometry imaging (MALDI-MSI). MALDI-MSI sample preparation requires thin fresh/frozen tissue sections coated with a chemical compound "matrix" to aid in the desorption/ionization of molecules by absorbing the energy delivered by a laser [1, 2]. The use of MALDI imaging for small molecules has recently been reviewed by Trim and Snel 2016 [3].

The instrument used in the work described in this chapter is a MALDI Synapt HDMS mass spectrometer (Waters Corporation, Manchester, UK). Although some details specific to this instrument are presented, many aspects of the analyses described in this chapter are applicable to other MS platforms and will serve as a useful guide to the techniques involved in MALDI-MSI analysis of large tissue sections.

Since its introduction in the mid-1990s [4, 5] MALDI-MSI has become a widespread analytical technique used to analyze small

Laura M. Cole (ed.), *Imaging Mass Spectrometry: Methods and Protocols*, Methods in Molecular Biology, vol. 1618, DOI 10.1007/978-1-4939-7051-3_15, © Springer Science+Business Media LLC 2017

molecules (lipids [6], drugs [7], etc.), and intact proteins and peptides [8–10]. To date, almost 200 reviews have been published on MALDI (Pubmed search "(MALDI [Title]) AND Review [Publication Type]"), with almost 30 reviews on MALDI imaging specifically (Pubmed search "(MALDI imaging[Title]) AND Review[Publication Type]").

A prominent reason for the uptake of MALDI-MSI as an analytical technique is its specificity and label-free nature. Using a mass spectrometer as the detector permits mapping of the spatial distribution of specific native analytes, whereas many other imaging modalities use either chemical stains [11], antibodies, or radiolabels [12], e.g., microscopy, autoradiography, etc. MALDI-MSI is a complementary imaging technique that provides additional information to these other imaging methods.

One area of significant interest is direct imaging of drug compounds and their metabolites [3, 7, 13, 14]. In the typical drug development pipeline understanding a novel compound's administration, distribution, metabolism, and excretion (ADME) properties is essential. This information is routinely established using a range of techniques, including high-performance liquid chromatography tandem mass spectrometry (HPLC MS/MS) and (quantitative) whole-body autoradiography ((Q)WBA).

HPLC-MS/MS is currently the gold standard [15] in quantitative analysis of known compounds due to its high sensitivity, specificity, and accuracy; however, spatial information is lost during sample preparation. Whole organ and some level of sub-organ quantitative analysis can be achieved by sample dissection prior to analysis [16]. Tissue distribution studies are commonly performed using (Q)WBA.

(Q)WBA requires the compound to be radiolabeled—commonly with ^{14}C or ^{3}H—making it an expensive and time-consuming process. Animals are dosed with the radiolabeled compound and euthanized after predetermined time points [12]. The animals are then snap-frozen, sectioned (this is common to both (Q)WBA and MALDI-MSI), and sections exposed to an imaging film. The original method of using X-ray film has been replaced by most laboratories by a phosphor imaging plate or radioluminography (Luckey 1975 [17]), which is then scanned by a phosphor image scanner to produce digital images of the radioactive label within the tissue section and in known standards allowing for quantitation. The whole process can take several weeks for the synthesis of a radiolabeled analog and subsequent (Q)WBA imaging. The reader interested in QWBA for drug distribution and quantitation is referred to a recent review by Solon 2015 [12] and for a comparative review of WBA, MALDI, and SIMS by Solon et al., 2010 [18].

Ultimately, MALDI-MSI may remove the need for developing radiolabeled analogs of novel compounds and has the potential to increase sample throughput; however, there are a few fundamental issues to overcome before this becomes a reality, including accurate analyte quantitation (recently reviewed by Ellis et al., 2014 [19]) and differences in analyte ionization efficiency [20, 21]. (Q)WBA measures the distribution of radiolabel but the specific identity of molecular species it is attached to is unknown (i.e., metabolites cannot necessarily be distinguished from the parent drug) [22], whereas MALDI imaging monitors the distribution of specific molecular ions (e.g., parent drug and/or metabolite [23–25]. High specificity is achieved by MS/MS analysis, however, the serial measurement of chemical species and the time taken for each transition limits the number that can be imaged per laser position/pixel. Alternatively, use can be made of mass accuracy and mass resolution to distinguish molecules of interest but this approach is limited by instrument performance characteristics and cannot distinguish isobaric species (e.g., structural isomers).

In a MALDI-MSI experiment the image resolution is determined by the size of the laser spot (excluding enhancements by oversampling). The smallest laser spot size, and hence highest image resolution available on a commercial MALDI system at the time of writing, is 5 μm [26]; however, such high lateral resolution is not commonly required for whole-body MALDI-MSI, and other instrument parameters such as mass accuracy, mass resolving power, and MS/MS capability tend to be more important [3].

Both MALDI-MSI and (Q)WBA use similar sectioning methods, which have changed very little over the years and still use a similar technique to that demonstrated by Ullberg (1954 [27], 1977 [28]).

2 Materials

All solvents should be high quality—preferably HPLC grade or higher; water should be ultrapure (18 MΩ cm at 25 °C).

2.1 Snap-Freezing and Blocking in Carboxymethyl Cellulose (CMC)

1. Hexane.
2. Pelleted dry ice.
3. Ice bath/bowl—a suitable size to fully submerge the whole animal and the frame without the risk of the hexane/dry ice overflowing.
4. Stereotaxic frame—a frame designed to hold the animal with the nose, spinal column, and base of tail in alignment during the freezing process (Fig. 1a).
5. Forceps—12″ surgical forceps or similar to remove the animal from the hexane/dry ice bath.
6. CMC—a 2% w/v solution in water prechilled to 4 °C.

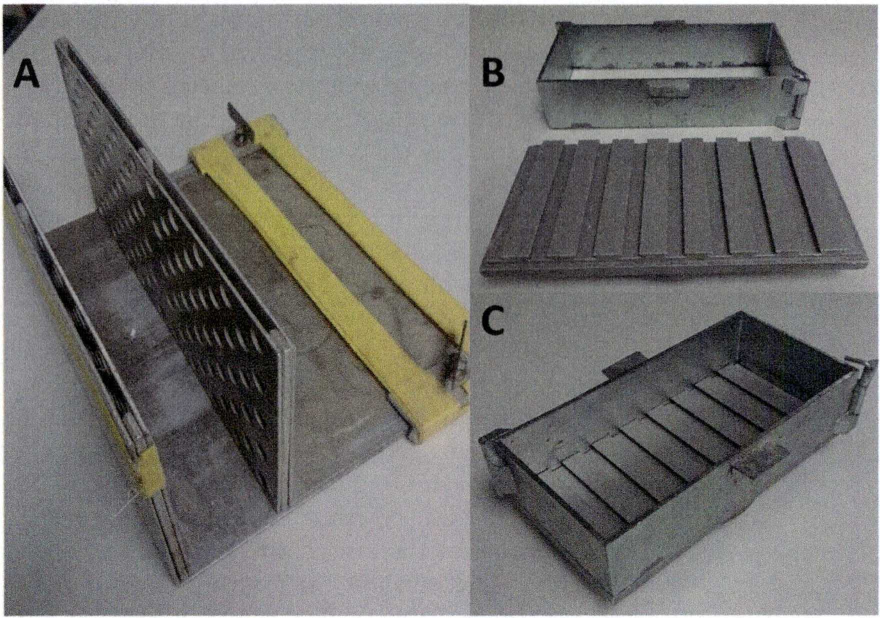

Fig. 1. (**a**) Stereotaxic frame used to freeze the carcass while maintaining spinal alignment. (N.B. the array of holes in each side permits the free flow of the hexane/dry ice freezing solution to the carcass). (**b**) The blocking mold and cryomacrotome stage, which combine to form the freezing mold (**c**)

7. Metal mold suited to the cryomacrotome large enough to fit the specimen. The mold shown in Fig. 1b, c combines the mold with the Leica stage in a single unit. Another option is a metal mold where the sample block is attached to the cryomacrotome stage with additional CMC after initial sample freezing.

2.2 Known Concentration Standards (If Analysis of Exogenous Compounds Is Required)

1. Standard curve/dilution series of the compound of interest in a suitable biological matrix, e.g., tissue homogenate. This should include a blank (tissue only) sample as well as four to eight samples at a concentration range from the lower limit of quantitation of the instrument to above the highest expected concentration.
2. CMC block with suitable sized holes drilled for each standard.

2.3 Cryo-Sectioning (See Note 1*)*

1. Specialized whole-body cryomacrotome similar to a Leica CM3600 XP (Leica microsystems, Wetzlar, Germany).
2. Macro-Tape-Transfer-System (Instrumedics Inc., St. Louis, MO, USA). If this system or similar is not available adhesive tape can be used and directly mounted to the target plate.
3. Small rubber hand roller.

2.4 Mounting

1. Clean, thin aluminum plate (similar in thickness to TLC plates)/MALDI target plate or glass slides precoated with the UV polymerising adhesive (part of the Macro-Tape-Transfer-System).
2. Sample holder suited to the mass spectrometer.
3. Double-sided tape.
4. Lyophilizer.

2.5 MALDI Matrix Coating

1. Sonic bath.
2. MALDI matrix 50 mL at 25 mg/mL αCHCA (α cyano-4-hydroxycinnamic acid) in 70:30 (v/v) ethanol:water (0.1% TFA) (*see* **Note 2**).
3. Pneumatic air spray gun with compressor (Iwata-Media Inc. OR, USA) or similar.
4. Heavy isotope labeled internal standard, if available, should be added to the matrix at a concentration that is similar to the expected analyte concentration and within the instrument's limits of detection (determined for the analyte of interest in suitable sample matrix).

2.6 MS Setup and Data Acquisition

1. Mass spectrometer with MALDI source. (Instrument setup varies between manufacturers: in the example given here a MALDI Synapt HDMS (Waters Corporation. Manchester, UK) Q-oaTOF MS is used.
2. A flat-bed scanner.
3. Positive ion calibration solution polyethylene glycol (PEG): 1 mg/mL PEG 600, 1000, 1500, and 2000 (average molecular weight), with 0.5 mg/mL sodium iodide in 2 mg/mL αCHCA in 50:50 water:acetonitrile (v:v).
4. Negative ion calibration solution PEG sulfate. Prepare a 1:10,000 (v:v) dilution of PEG sulphate in 2 mg/mL αCHCA in 50:50 water:acetonitrile (v:v).

2.7 Data Analysis

1. Depending on instrument manufacturer data analysis may vary. Most data can be converted into files readable by BioMap (Novartis, Basel, Switzerland) [29]. BioMap is freely available software suitable for MALDI image analysis and can be downloaded from www.maldi-msi.org.

3 Methods

3.1 Snap-Freezing (45 min)

1. Prepare the hexane/dry ice bath by half-filling a large rectangular bowl with hexane (size depends on sample; bowl must be large enough to contain hexane/dry ice solution and allow the sample to be fully submerged), then carefully adding dry ice

pellets to the hexane (**CARE**, *see* **Note 3**). The amount of dry ice pellets required will vary depending on size of the bowl and ambient temperature, etc. Dry ice should be added until the solvent is cold and some pellets of dry ice remain floating in the hexane.

2. The euthanized rodent is placed in a frame to ensure alignment before freezing (Fig. 1a). If a frame is not available, gently pulling the head and tail in opposite directions to get the spinal column in line using forceps will achieve a similar outcome. Pinning to a cork board can also be used if required.
3. While maintaining the correct alignment submerge the rodent in the hexane/dry ice bath until completely covered. Fully freeze the sample; the time required varies depending on sample size. For instance, a typical male Sprague-Dawley rat (ca. 260 g) should be frozen for 30 min.

3.2 Blocking in CMC (30 min)

1. The freezing mold with stage (Fig. 1b, c) is precooled by placing on a bed of dry ice; a small amount of prechilled CMC is added to the bottom of the mold. The frozen carcass is then placed into the center of the container and the surrounding area is filled with prechilled (4 °C) 2% CMC (w/v). Ensure the container is at least three-quarters submerged and leave in the hexane/dry ice bath for 30 min or until fully frozen (the dry ice will evaporate over time and will need to be topped up as required). Once frozen the sample can be removed from the container and stored in a sealed bag at −80 °C until required.

3.3 Addition of Standards (Quantitation)

1. Relative quantitation of exogenous compounds (e.g., pharmaceutical agents) can be achieved using blood or tissue homogenates spiked with the compound of interest at concentrations covering the expected range. This standard curve can be added to a CMC block by drilling holes in the block and filling these with spiked homogenates and freezing in the hexane dry ice bath. (N.B. at this stage the sample can be stored in a sealed bag at −80 °C until required).

3.4 Preparation of Mounting Media

1. Ensure the surface of the aluminum mounting plate is clean and flat (*see* **Note 4**).
2. Coat the mounting plates with the UV curable adhesive supplied with the Macro-Tape-Transfer-System (Instrumedics Inc., St. Louis, MO, USA) following the manufacturer's instructions. Skip this step if using direct tape mounting without a tape transfer system.

3.5 Cryo-Sectioning (All Equipment and All Steps Are Completed Inside the Cryotome Chamber)

1. For whole-body sectioning a specialized cryomacrotome is required (e.g., Leica CM3600 XP); smaller samples can be sectioned using a standard cryomicrotome.
2. Securely mount the block into the holder in the cryomacrotome, ensuring the correct orientation for sectioning (sagittal/coronal).
3. Using a fresh blade (care: the blade is extremely sharp) for each sample start to trim in the sectioning using large steps 50–100 μm until the required tissue depth is reached.
4. Set the cryomacrotome to 30 μm section thickness; cut and discard the next section.
5. Apply the adhesive tape to the top of the sample block and ensure it is firmly adhered using a rubber or foam hand roller.
6. Slowly cut the section while applying slight pressure a few millimeters in front of the cutting blade to prevent the tape lifting off the tissue. When cutting the section gently lift the tape with the attached section as it is being cut, ensuring not to pull the tape because the section can easily be damaged.
7. Cutting a second serial section for histological staining can be very useful for comparison with the MALDI image and will provide cellular information. (This section should be mounted to a glass slide for staining.)
8. Repeat **steps 3–7** for all the tissue levels required.
9. If using direct tape mounting stick the sample tissue side up directly on the target plate using double-sided tape (skip **steps 10–12**).
10. Place the tissue section face down onto the precoated mounting medium, ensuring good adhesion using a hand roller.
11. Following the manufacturer's instructions polymerize the adhesive using a pulse of long wave UV light.
12. Carefully remove the tape from the tissue section, leaving it firmly attached to the mounting medium.

3.6 Freeze Drying/ Lyophilizing

1. Remove the section(s) from the cryostat chamber along with a metal block at least the size of your sample and place directly in the freeze-drier chamber. The metal block prevents the section(s) from thawing between the transfer and the start of the freeze drying process (the very thin tissue section would rapidly thaw if not kept on a frozen metal block).
2. Immediately evacuate the freeze dryer and dry the section for 45 min. (Fr4eeze-dried sections can be stored in an air-tight container or bag at −80 °C until required. However, when defrosting before use ensure the container/bag remains sealed until the sample has reached room temperature to prevent condensation forming on the tissue section).

3.7 MALDI Matrix Coating

1. Prepare the MALDI matrix solution (Subheading 2.5, **item 2**) and sonicate for 15 min before use (*see* **Note 5**).
2. Skip **steps 3** and **4** if the section is being directly mounted onto target or glass slide.
3. On a flat surface, using a scalpel and straight edge, carefully cut the sample and backing material to the correct size to fit the target plate (for the MALDI Synapt 50 mm (w) × 40 mm (h)).
4. Stick double-sided tape to the target plate ensuring good adhesion, then peel off the protective backing, and stick the mounted tissue section to the target plate (tissue section up). Firmly press the sample backing, making sure not to damage the sample and ensuring it is firmly held in position without significant variations in sample height due to creasing.
5. If used, the spiked standards should be added to the target plate using the same technique.
6. Tape the target plate on a suitable support (the inside of a large cardboard box with the sides cut away works well). Place the whole assembly in a fume hood (Fig. 2).
7. Using a gravity-fed pneumatic air spray gun set to 40 psi air pressure, coat the sample with the full volume of matrix solution by spraying in a horizontal sweeping motion, and keeping the airbrush 25 cm away from the sample (*see* **Note 6**). To ensure the sample does not get "wet" perform the sweeping motion across the sample in two passes (over and back). The flow of matrix solution to the airbrush should be started and stopped while off the sample. Allow 30–45 s between each coat for the matrix to dry on the sample.
8. Once all matrix has been deposited on the sample allow to fully dry for at least 5 min (*see* **Note 7**).

3.8 Instrument Mass Calibration

1. Calibrate instrument according to the manufacturer's instructions. A brief overview of the steps involved to calibrate a MALDI Synapt HDMS mass spectrometer is given below.
2. Prepare the calibration solution (Subheading 2.6, **items 3** and **4**) (*see* **Note 8**).
3. Spot 1 μL droplets onto a clean target plate and allow to air dry.
4. Acquire MS only data over the required mass range and calibrate the instrument to within 3 ppm mass error.

3.9 MALDI Imaging Data Acquisition

1. The following steps outline the image acquisition method using a MALDI Synapt HDMS mass spectrometer.
2. Using a flat-bed scanner obtain an image of the tissue section mounted on the target using a minimum 600 dpi (higher resolution generates higher quality images). This is required for setting up the image area on the mass spectrometer and can also be used to overlay the MALDI image.

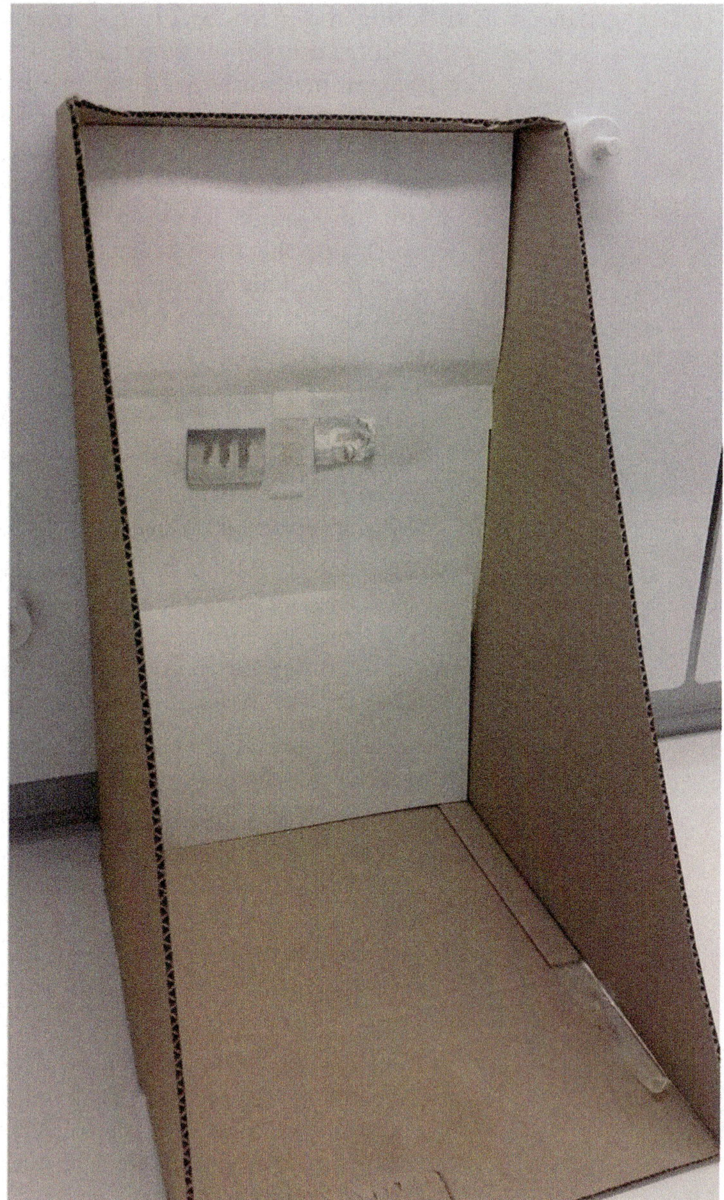

Fig. 2. Tissue sections mounted on a cardboard support inside a fume cupboard awaiting spray coating with matrix

3. Place the sample target into the calibrated mass spectrometer.
4. Load the scanned image into MALDI pattern creator (Waters Corporation, Manchester UK) (*see* **Note 9**). Using the four corners of the target plate as fiducials, an area encompassing the whole tissue section can be selected and used to generate region pattern data (area to image) for image acquisition.
5. Set the image resolution to a step size of 300 μm in both the *x* and *y* directions (*see* **Note 10**).

6. Using the method editor software in MassLynx v4.1 (Waters Corporation, Manchester, UK) set the method parameters: for maximum sensitivity a laser repetition rate of 200 Hz and laser energy of 250 a.u. is normally optimal. Acquisition duration of 1–2 s per pixel is commonly used. The duration required for each laser position can be determined by manually firing the laser at a single locus on the spiked tissue homogenate and observing the time taken before the signal intensity rapidly decreases. This duration should be set as the scan time per pixel for the image acquisition.
7. If using the spiked tissue standards the MS/MS conditions with the highest sensitivity and specificity for the compounds of interest can be established. Another technique, which can also be used and set up using the spiked tissue standards, is ion mobility separation (IMS) either with or without MS/MS fragmentation (*see* **Note 11**).
8. Alternatively MS only imaging can be performed: this has a lower specificity compared to MS/MS but captures information from the entire mass range. This permits distribution analysis of unknown compounds, which can later be confirmed using an MS/MS approach if required.
9. Acquire the MALDI image using the optimum conditions established above (**steps 6–8**).
10. Following acquisition, data require conversion to a format suitable for image visualization: several vendor/commercial and freeware options are available for image analysis. The platform used by a large number of imaging laboratories using various instruments is BioMap. BioMap is freely available software that can be downloaded from www.maldi-msi.org.
11. Using MALDI image converter (Waters Corporation, Manchester, UK) the acquired data files can be converted from *.raw* format to *analyze* format using exact pixel coordinates and the desired *m/z* bin size. A bin size of 0.03 Da is a good compromise between file size and maintaining the quality of the data acquired.

3.10 Image Analysis Using BioMap

1. Detailed instructions in the use of BioMap can be found at www.maldi-msi.org.
2. Open BioMap and import the MSI image file (.img) generated in Subheading 3.9, **step 11**.
3. Selected ion intensity maps can be generated in BioMap showing the relative intensity of a specific ion in the tissue section (Fig. 3).
4. Ion suppression is a major issue with whole-body imaging (as it can be with all MALDI-MSI experiments); therefore, normalization of the data is required to obtain a true representation of the distribution of a compound of interest.

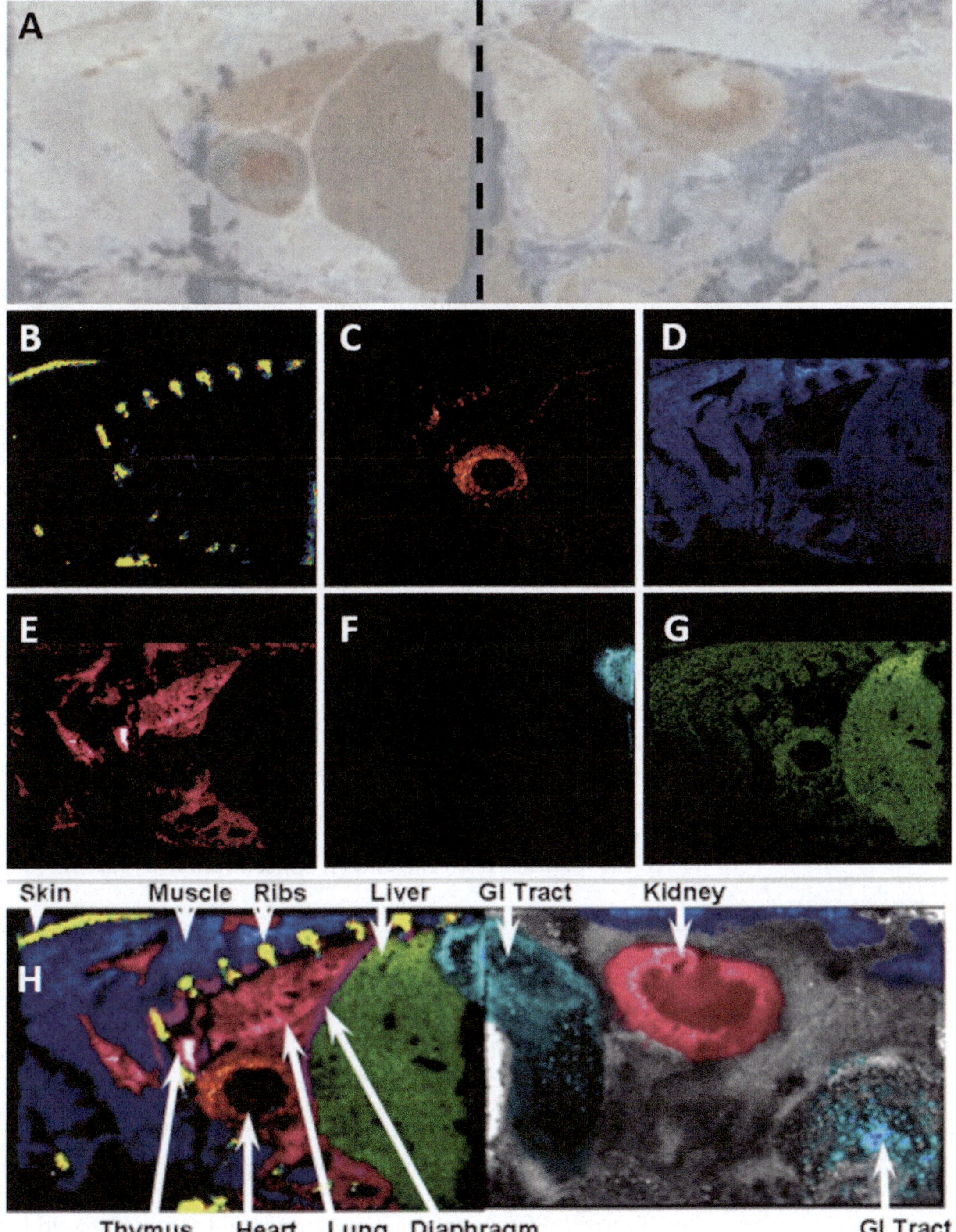

Fig. 3. (**a**) Scanned image of the thoracic and abdominal region of a 250 g Sprague-Dawley rat 30 μm frozen section mounted on aluminum plates prior to matrix coating; (**b–g**) MALDI-MSI of the same tissue section. A different ion map is shown in each panel using a different color scale for each ion; (**h**) an overlay of images (**b**)–(**g**) with the addition of the second plate

5. Select an ion for normalization. Currently, the optimum method uses a spiked heavy isotope labeled version of the ion of interest in the matrix. If an internal standard is not available a known matrix ion can be used, e.g., for αCHCA the molecular ion *m/z* 190.0504 or molecular ion after a loss of water $[M-H_2O+H]^+$ *m/z* 172.0399. Divide the intensity of the ion of interest by the intensity of the internal standard/matrix ion at the same pixel throughout the image using the intensity calculation function in BioMap.
6. Normalization to a matrix ion can be difficult when performing MS/MS experiments using a narrow precursor window that is devoid of matrix ions, so care in the interpretation of the relative concentration of the analyte must be taken.
7. After normalization images showing the distribution of the ions of interest can be coregistered with the scanned image of the tissue section obtained prior to imaging or an H&E-stained serial section.
8. Due to the relatively low image resolution of whole-body imaging, coregistration using tissue regions or outlines is generally sufficiently accurate. This is achieved by rescaling the scanned image to match the MALDI image. Obtain the dimension of the two images and edit the header file of the scanned image by a scaling factor.
9. Extracted ion images can then be copied and overlaid on the scanned image.
10. Relative quantitation of the ion of interest can be obtained after normalization using the spiked tissue homogenates (*see* **Note 12**). The relative intensity of each pixel can be compared to the intensity from the known concentration standards.

4 Notes

1. When working inside the cryomacrotome ensure all equipment, including tape, is kept in the cryomacrotome chamber and is not at room temperature.
2. In the example given here, αCHCA was used as the matrix: this is generally a good matrix to use with the MALDI Synapt HDMS mass spectrometer for several applications. If standards of the compounds of interest are available, an initial assessment of several different matrix compounds and solvents should be made. The matrix/solvent that results in the highest sensitivity should be used. Interference in the low mass range between the compound of interest and matrix ions/clusters should be considered during matrix choice.

3. Dry ice can cause burns and needs to be handled with care and suitable PPE needs to be worn, including laboratory coat, safety glasses, and gloves. It should always be used in a well-ventilated area or fume hood.
4. Other mounting material can be used depending on the source design of the mass spectrometer. ITO-coated glass slides are a good alternative option as well as the manufacturer's MALDI target plates. The same steps can be used with any of these other materials.
5. When relative quantification of a specific compound of interest is required, the addition of a heavy isotopically labeled analog of the compound can be spiked into the matrix solution prior to spray coating. This can be used to normalize the tissue-specific suppression effects, which are frequently observed.
6. Extreme care is required during the matrix-coating process: if the sample becomes too wet during the application stage the risk of analyte spreading/redistribution within the sample is very high thus invalidating any assessment of analyte distribution within the tissue; over-wetting the section can also create redistribution of analytes resulting in a characteristic high ion intensity region surrounding the tissue.
7. Matrix coating should be assessed throughout the application process, and the sprayer distance from the sample adjusted if required. Matrix volume and concentration should also be adjusted if required to create a uniform matrix coating with small uniform matrix crystals. Matrix coating can be checked using a dissection microscope.
8. The selection of PEG molecular weights can be modified to cover other ranges of the mass spectrum to ensure the mass range of the analytes is suitably covered.
9. The latest version of software available from Waters Corporation for image acquisition and data processing/visualization is HDI v1.4: this allows data acquisition and coregistration of images as well as data processing and statistical analysis within the same software platform. This version also permits analysis of MALDI IMS MSI data.
10. Image resolution required for whole-body imaging is generally lower than that used for small tissue samples. Between 150 and 1000 μm step sizes are used. This is because every time the step size is halved the number of acquisition points is quadrupled. This results in high-resolution images that are very time-consuming and produce exceptionally large data sets.
11. MALDI-IMS-MSI is currently only commercially available on very few instruments. IMS provides an orthogonal means of analyte separation. This technique can be used in MS or MS/

MS acquisition mode, enhancing the specificity of the analysis if interfering ions are present and conventional MS/MS analysis does not resolve the issue.

12. Absolute quantitation is not possible (at the time of writing) using MALDI-MSI, only relative quantitation. This is due to many factors, including the heterogeneity of tissues and matrix coating, the ionization efficiency and ion suppression effects. Many of these aspects can be controlled for by careful experimental design and sample preparation methods and the use of suitable standards.

Acknowledgments

I wish to thank Claire Henson for her invaluable technical advice, and Dr Marten Snel for technical discussions and guidance over many years.

References

1. Dreisewerd K (2003) The desorption process in MALDI. Chem Rev 103:395–425
2. Jaskolla TW, Karas M (2011) Compelling evidence for lucky survivor and gas phase protonation: the unified MALDI analyte protonation mechanism. J Am Soc Mass Spectrom 22:976–988
3. Trim PJ, Snel MF (2016) Small molecule MS imaging: current technologies and future challenges. Methods 104:127–141
4. Caprioli RM, Farmer TB, Gile J (1997) Molecular imaging of biological samples: localization of peptides and proteins using MALDI-TOF MS. Anal Chem 69:4751–4760
5. Spengler B, Hubert M, Kaufmann R. MALDI ion imaging with a new scanning UV-laser microprobe. In: Proceedings of the 42nd ASMS Conference on Mass Spectrometry and Allied Topics, May 29–June 3, 1994, Chicago, IL, ThP 1041, Thursday
6. Sparvero LJ, Amoscato AA, Dixon CE, Long JB, Kockanek PM, Pitt BR, Bayir H, Kagan VE (2012) Mapping of phospholipids by MALDI imaging (MALDI-MSI): realities and expectations. Chem Phys Lipids 165(5):545–562
7. Castellino S, Groseclose MR, Wagner D (2011) MALDI imaging mass spectrometry: bridging biology and chemistry in drug development. Bioanalysis 3(21):2427–2441
8. Djidja M-C, Chang J, Hadjiprocopis A, Schmich F, Sinclair J, Mršnik M, Schoof EM, Barker HE, Linding R, Jørgensen C, Erler JT (2014) Identification of hypoxia-regulated proteins using MALDI-mass spectrometry imaging combined with quantitative proteomics. J Proteome Res 13:2297–2313
9. Hanrieder J, Malmberg P, Ewing AG (2015) Spatial neuroproteomics using imaging mass spectrometry. Biochim Biophys Acta 1854: 718–731
10. Andersson M, Andren P, Caprioli RM (2010) MALDI Imaging and profiling mass spectrometry in neuroproteomics. In: Alzate O (ed) Neuroproteomics. CRC Press/Taylor & Francis, Boca Raton, FL. Chapter 7
11. JKC C (2014) The wonderful colors of the hematoxylin-eosin stain in diagnostic surgical pathology. Int J Surg Pathol 22(1):12–32
12. Solon EG (2015) Autoradiography techniques and quantification of drug distribution. Cell Tissue Res 360:87–107
13. Prideaux B, Stoeckli M (2012) Mass spectrometry imaging for drug distribution studies. J Proteomics 75:4999–5013
14. Greer T, Sturm R, Li L (2011) Mass spectrometry imaging for drugs and metabolites. J Proteomics 74(12):2617–2631
15. Ramanathan R, Jemal M, Ramagiri S, Xia Y-Q, Humpreys WG, Olah T, Korfmacher WA (2011) It is time for a paradigm shift in drug discovery bioanalysis: from SRM to HRMS. J Mass Spectrom 46:595–601
16. Jones DC, Duvauchelle C, Ikegami A, Olsen CM, Lau SS, de la Torre R, Monks TJ (2005)

Serotonergic neurotoxic metabolites of ecstasy identified in rat brain. J Pharmacol Exp Ther 313(1):422–431

17. Luckey G (1975) US Patent 3:859,527
18. Solon EG, Schweitzer A, Stoeckli M, Prideaux B (2010) Autoradiography, MALDI-MS, and SIMS-MS imaging in pharmaceutical discovery and development. AAPS J 12(1):11–26
19. Ellis SR, Bruinen AL, Heeren RMA (2014) A critical evaluation of the current state-of-the-art in quantitative imaging mass spectrometry. Anal Bioanal Chem 406:1275–1289
20. Tomlinson L, Fuchser J, Fütterer A, Baumert M, Hassall DG, West A, Marshall PS (2014) Using a single, high mass resolution mass spectrometry platform to investigate ion suppression effects observed during tissue imaging. Rapid Commun Mass Spectrom 28:995–1003
21. Pirman DA, Kiss A, Heeren RMA, Yost RA (2013) Identifying tissue-specific signal variation in MALDI mass spectrometric imaging by use of an internal standard. Anal Chem 85:1090–1096
22. RJA G, Nilsson A, Mackay CL, Swales JG, Johansson MK, Billger M, Andrén PE, Iverson SL (2016) Exemplifying the screening power of mass spectrometry imaging over label-based technologies for simultaneous monitoring of drug and metabolite distributions in tissue section. J Biomol Screen 21(2):187–193
23. Trim PJ, Henson CM, Avery JL, McEwen A, Snel MF, Claude E, Marshall PS, West A, Princivalle AP, Clench MR (2008) Matrix-assisted laser desorption/ionisation-ion mobility separation-mass spectrometry imaging of vinblastine in whole body tissue sections. Anal Chem 80(22):8628–8634
24. Jones EE, Gao P, Smith CD, Norris JS, Drake RR (2015) Tissue biomarkers of drug efficacy: case studies using a MALDI-MSI workflow. Bioanalysis 7(20):2611–2619
25. Prideaux B, Dartois V, Staab D, Weiner DM, Goh A, Via LE, Barry CE 3rd, Stoeckli M (2011) High-sensitivity MALDI-MRM-MS imaging of moxifloxacin distribution in tuberculosis-infected rabbit lungs and granulomatous lesions. Anal Chem 15(83):2112–2118
26. Potočnik NO, Porta T, Becker M, Heeren RMA, Ellis SR (2015) Use of advantageous, volatile matrices enabled by next-generation high-speed matrix-assisted laser desorption/ionisation time-of-flight imaging employing a scanning laser beam. Rapid Commun Mass Spectrom 29:2195–2203
27. Ullberg S (1954) Studies on the distribution and fate of 35S-labelled benzylpenicillin in the body. Acta Radiol Suppl 118:1–110
28. Ullberg S (1977) The technique of whole-body autoradiography: cryosectioning of large specimens. In: Elvefeldt O (ed) Special issue on whole-body autoradiography. Science Tools: LKB Instrument Journal. LKB-produkter, Bromma
29. Stoeckli M, Staab D, Staufenbiel M, Wiederhold K-H, Signor L (2002) Molecular imaging of amyloid β peptides in mouse brain sections using mass spectrometry. Anal Biochem 311(1):33–39

Chapter 16

The Future in Disease Models for Mass Spectrometry Imaging, Ethical Issues, and the Way Forward

Rebecca E. Day and Ieva Palubeckaite

Abstract

Mass Spectrometry Imaging (MSI) has evolved into a valuable tool for research into and the diagnosis of disease pathology. The ability to perform multiplex analysis of a wide range of molecules (e.g., proteins, lipids, and metabolites) simultaneously per tissue section while retaining the histological structure of the sample allows molecular information and tissue morphology to be correlated, thus increasing our understanding of a particular disease. Further development of MSI is required to improve suitability to the alternative models available, so that the combined approach can successfully provide the information required in disease characterization and prevention. MSI has been shown to be capable of providing spatiomolecular information in tumor spheroids, living skin equivalents, and ex vivo human tissues. Due to a considerable interest and scientific effort there are many more designed alternative disease models available which would benefit from the information MSI could provide.

Key words Multiplex analysis, Disease models, Tumor spheroids, Living skin equivalents, Spatiomolecular

1 Mass Spectrometry Imaging of Disease

Biological tissue of both human and animal origin is vastly complex; the spatial and temporal variation in the molecular processes of disease requires advances in experimental techniques and instrumentation in order to further elucidate the pathological processes of disease. Mass Spectrometry Imaging (MSI) has evolved into a valuable tool for research into and the diagnosis of disease pathology. The ability to perform multiplex analysis of a wide range of molecules (e.g., proteins, lipids, and metabolites) simultaneously per tissue section while retaining the histological structure of the sample allows molecular information and tissue morphology to be correlated, thus increasing our understanding of a particular disease. There are several advantages to using MSI for the analysis of tissues, especially with proteomic studies. Immunohistochemistry, for example, is a robust and popular technique for the investigation of protein expression and localization

Laura M. Cole (ed.), *Imaging Mass Spectrometry: Methods and Protocols*, Methods in Molecular Biology, vol. 1618, DOI 10.1007/978-1-4939-7051-3_16, © Springer Science+Business Media LLC 2017

within tissues. However, much like many conventional protein analysis techniques, it involves lengthy protocols for the extraction and detection of target proteins. Furthermore, although this technique can provide spatial information, it is majorly limited in the amount of data that can be acquired from each experiment and the target protein must be known in advance in order to select the appropriate antibody for staining. On the other hand, MSI can be used for the discovery of proteins related to a specific disease or treatment with no target-specific reagents needed; in addition to this, it is highly specific and hundreds if not thousands of peptides can be detected simultaneously from one experiment on one single tissue section. Following MSI, the tissue section that has been imaged can be removed from the mass spectrometer and subjected to histological staining, such as Hematoxylin and Eosin staining, to provide architectural information about the exact piece of tissue which has been imaged.

MSI is commonly used in various modalities to study the pathogenicity of disease progression or tissue response to treatments; matrix-assisted laser desorption ionization MSI (MALDI-MSI) was used to observe the spatial distribution of peptides within a mouse fibrosarcoma model following treatment with the tubulin-binding tumor vascular disrupting agent, combretastatin A-4-phosphate (CA-4-P) [1]. Alterations in peptide distribution as a result of an increase in hemoglobin following treatment with CA-4-P were observed within the MALDI-MSI images. Additionally, MSI is also showing its potential to be applied to disease diagnosis and classification. This is particularly useful in diseases such as breast cancer where the heterogenic nature means patients often have very different survival rates due to tumor phenotypes affecting the metabolic response to therapeutic agents [2, 3]. Guenther and colleagues performed a study using Desorption Electrospray Ionization MSI (DESI-MSI) in order to diagnose and evaluate the metabolic phenotype of human breast tumor biopsies [4]. DESI-MSI was able to achieve 98.2% accuracy in breast cancer diagnosis along with a correlation between the metabolomics profile and grade of the tumors with hormone receptor status. Human breast cancer tumor type and grade has also recently been analyzed by Airflow-assisted ionization MSI (AFAI-MSI) based on the lipid profiles of invasive ductal carcinomas and in situ ductal carcinomas [5]. AFAI-MSI does not require any sample pretreatment or labeling and analysis results in the production of a multi-color map illustrating the spatial distribution and intensities of molecules within a tissue. MSI research has now entered the frontier of whole-body imaging allowing a multi-modality approach to biomedical imaging. MALDI-MSI was used to detect inflammatory protein responses in rats infected with *Staphylococcus aureus* [6]. The specificity of MALDI-MS was used in conjunction with high-resolution images produced by magnetic resonance imaging to produce an image obtaining molecular and structural information on the distribution of immune response proteins throughout the entire rat body following systemic infection.

2 Tissue Sample Preparation: Fresh Frozen

When tissue is harvested for use in scientific research applications, specifically MSI, it is vital that such specimens are handled correctly in order to preserve tissue viability. It is imperative that minimal time passes from the harvest event to freezing the tissue as ischemia, nutrient starvation, and temperature will all contribute to enzymatic degradation within the tissue which will lead to poor sample quality for MS analysis, particularly for proteomic and metabolomic analysis. Snap freezing tissue in isopentane cooled by liquid nitrogen [7] or floating the specimen on liquid nitrogen is advised as immersing the tissue within liquid nitrogen can induce tearing and cracking [8]. The main disadvantage of fresh frozen tissue is the limited lifespan of the samples and demand on freezer storage space, if samples are not stored in an embedding medium they will begin to dry out over time making sectioning and subsequent analysis difficult. However, this method of sample preparation for MSI is often favorable due to the reduced washing steps therefore retaining small molecules, drugs, and lipids for the detection by MSI, especially important when performing drug distribution studies.

3 Tissue Sample Preparation: Formalin-Fixed Paraffin Embedded

Formalin fixation followed by embedding in paraffin wax is a widely used tissue preservation and storage technique that allows the long-term storage of tissues at room temperature. Formalin-fixed paraffin-embedded (FFPE) tissue is first dehydrated in graded ethanol and cleared in Xylene before fixation in 10% neutral buffered formalin; this step induces the formation of protein crosslinks within the tissue [9]. The specimen is then embedded within paraffin wax and is stored permanently like this ready for sectioning.

There is a vast library of available FFPE tissue samples; this provides scientists with the ability to acquire data sets from large patient cohorts, very important when obtaining statistically relevant data from the diverse human population pool. FFPE tissue can be archived for many years in hospitals and universities, along with data from the patient the specimen originated from. These tissue archives allow scientist's access to large cohorts of samples for investigation of specific diseases. As technology advances samples can be reanalyzed providing a rich source of new information. A study using an FFPE biopsy from the spleen of a patient who died from tuberculosis and secondary amyloidosis in 1899 was analyzed revealing the presence of peptides relating to serum amyloid [10]. Another advantage of FFPE tissue is its suitability to be used for the creation of tissue microarrays (TMAs). TMAs are formed from needle core biopsies from entire patient cohorts (often hundreds) that are placed into one paraffin block. This paraffin block will then be sectioned and subjected to MSI and pathological examination allowing the acquisition of data from an entire patient cohort simultaneously. TMAs were originally developed as a higher throughput method of performing molecular and pathological analysis of tissue specimens, such as protein expression studies using IHC [11, 12]. TMAs have since been shown to provide a valuable method for protein analysis by MSI; a study using a TMA comprised of 112 needle core biopsies from lung tumor patients showed that incorporating antigen retrieval, on-tissue enzymatic digestion, and MSI proteomic analysis combined with histochemical analysis could provide the ability for classification of lung tumor biopsies [13]. Another study using MALDI-MSI with ion-mobility separation in combination with principal component analysis (PCA) showed significant differentiation between tumor types due to the distinct peptide profiles and distribution within images of a pancreatic carcinoma TMA [14].

Although FFPE tissue is described by some as the "gold standard" for histopathological analysis of tissue [15], especially proteomic analysis; challenges are still faced with the analysis of lipids and exogenous pharmaceuticals by MSI. The processing of FFPE tissue specimens includes many washing steps and de-waxing that washes away drug compounds and strips lipids from the tissues. The analysis of lipids in biological tissues is of great importance as they are often shown to play a key role in the development and progression of various diseases [16]. There is currently very little literature on the analysis of small molecules and metabolites within FFPE tissue. However, a recent study used high-resolution MALDI-FT-ICR MSI for in-situ imaging of metabolites in FFPE human tissue. Initially, they compared spectral signals from both FF and FFPE between m/z 50 and 1000 m/z revealing a high proportion of overlapping peaks and interestingly peaks from FFPE tissue were more intense between m/z 50 and 400. Signal intensity in the 600–1000 m/z range was more intense in the FF samples, perhaps because a large proportion of peaks within this mass range are lipids. Following this, using FFPE tissue, they were able to distinguish between tumor tissue and healthy mucosa, chromophobe renal cell carcinoma and renal oncocytoma, a distinction that is difficult to make using just histology and finally a prognostic marker for oesophageal adenocarcinoma was identified [15]. This data shows promise for MSI of small molecules within FFPE tissue and as it is known that small molecules and cellular components can be trapped within the meshwork of protein cross-links as a result of formalin fixation; further method development into this application would be worthwhile [17].

4 Human vs. Animal Disease Models

The analysis of human biological tissue samples by MSI currently poses more challenges and limitations due to the level of genetic diversity of human tissues and diseases compared with that of animal models; discoveries made from studies using animal models may not accurately represent the biological processes within human systems [18]. The heterogeneity, disease aetiology, and diverse environmental factors associated with human samples mean that a wide variation in data is seen with human samples, even those affected with the same disease [19]. Additionally, biopsies are often either difficult to obtain or are not suitable for the study performed due to ethical considerations; therefore, many tissue samples derive from animal origin. It is well known that the animal model is prominently integrated into scientific research. In the study of disease, an animal model is often used as a surrogate of a human subject. These studies often involve creating disorders in the animal by chemical, genetic, or surgical means in order to mimic the disease of interest or accurate animal response to treatment. However, the animal model has received an abundance of criticism for being inappropriate for use to test human disease [20]. Various animal models of disease have been shown unable to recreate the complexity of the human genome, physiology, and environment. The effect is observed in both disease characterization and treatment studies, and is becoming particularly prominent in the drug discovery field. A group of leading pharmaceutical companies have released publications on the failure of animal model-derived results to translate to clinical studies and are calling for improvement of the current models and further self-assessment of the effectiveness of their use [21]. Several studies have shown the predictive ability of prevalent models to be limited, when taking the most commonly observed adverse effects into consideration [22, 23]. Therefore, the general consensus is that improvement is required in the way that animal research is performed, and the ability of this model to sufficiently recapitulate human clinical response should be closely monitored.

5 Alternative Models

Presently, the majority of the scientific community are aiming to replace, reduce, and refine the use of animals in research and testing (the 3Rs). One of the ways these goals can be achieved is by the design of alternative disease models. Alternative disease models include improved in silico modeling (computer simulation), ex vivo (outside of the organism) tissue models cultured outside of a human donor, and enhanced in vitro three-dimensional (3D) cell culture models.

For instance, alternative disease modeling has been embraced by skin disease scientists. Many ex vivo and reconstructed human skin equivalents are now available for skin penetration, characterization, and metabolism studies. Several studies have also developed use of skin disease phenotype models such as peeling skin disease (PSD) and non-melanoma skin cancer (NMSC) that were used by Alnasif and colleagues to study effect of nanotransporter penetration of normal, damaged, and diseased human skin [24]. Three-dimensional cell cultures have likewise succeeded in the design of Alzheimer's disease models. Currently established mouse models with familiar Alzheimer's disease (FAD) mutations exhibit only a fraction of the key pathological events of the disease. Using a neural stem-cell-derived 3D culture system produces a model capable of recapitulating the amyloid-β and tau pathology present in vivo [25]. Alternative models of disease are becoming more prevalent due to their human origin, cheaper cost, robustness, higher throughput, and ethical advantages. Consequently, there is a necessity to develop the optimal methods required to investigate these models.

6 MS of Alternative Models

Ex vivo and in vitro alternative models are already commonly used in combination with mass spectrometry, in a nonimaging modality, in order to create an effective method for the characterization of the model and its translatability to in vivo results. Liquid-chromatography mass spectrometry (LC-MS) has recently been applied to the analysis of the metabolome of 3D constructed cultures in order to study the metabolic reprogramming that occurs in cancer [26]. The 3D cancer culture was developed specifically with a rapid unrolling of the bio composite layers for hypoxic gradient investigation. Using their unique TRACER system and mass spectrometry they were able to observe changes in the metabolome consistent with known hypoxia mechanisms, such as the glycolytic switch and decreased levels of glutathione. They were also able to identify cellular adaptations to hypoxia gradients in the form of up/downregulated metabolites with higher hypoxia levels. LC-MS was also used to characterize the processed human cardiac extracellular matrix protein composition, in an attempt to improve a decellularized human cardiac tissue model. The method even managed to recognize residual protein material "contamination" which was not previously apparent with conventional assays for residual DNA and ECM proteins [27].

7 MSI of Alternative Disease Models

The use of MSI for the analysis of alternative disease models has several potential advantages. The main two advantages are the potential for untargeted de novo discovery, and the capability to observe multiple analytes of interest in a single experiment. Unlike LC-MS, the method that has mostly been used in the analysis of alternative models, MSI, is also able to preserve topographical information, requiring no homogenization of the sample, and frequently less sample preparation is involved. Many MSI methods are also nondestructive which may lead to the sample being used again, commonly for a histological stain. In light of these advantages MSI will become a powerful, regularly used technique for the investigation of alternative disease models.

Compared to the amount of research conducted every year on ex vivo and 3D cell culture models, only a miniscule fraction have utilized MSI as a method of validating and facilitating the study of disease. A number of research groups have begun work regarding this. For example, there is currently a vast amount of interest in three-dimensionally cultured cancer tumor spheroids, which are being used as in vitro models of cancer for characterization and drug dynamics studies. The first proof-of-principal publication on the use of MSI with these models, by Li et al., proved that a MALDI-MSI workflow could be used to examine the changes in protein and peptide distributions within cancer spheroids in an unbiased fashion, suitable for de novo discovery [28]. Recently, the same group produced a dynamic flow version of the cancer spheroid model, enabling the reconstruction of more representative conditions with the use of a fluidic device. The platform was constructed to assess drug penetration and metabolism and allow dynamic dosing of the chemotherapeutic drug irinotecan. Drug penetration into the spheroids was observed, as well as the distribution of its active metabolite SN-38 inside the necrotic, core region [29]. MALDI-MSI has also been applied to the study of tissue-engineered living skin equivalent (LSE) models. An established LSE named Labskin, consisting of a fibrin-based 3D scaffold with defined dermal regions, was examined using MALDI-MSI during its maturation as well as its response to emollient cream application. Alternative models like Labskin may be required in the cosmetics industry since the decision of the European Commission to ban animal testing in 2009 [30]. The developed workflow was able to recognize specific species for each type of emollient cream applied to the skin equivalents, as well as several endogenous lipid species previously shown to be associated with skin maturation [31]. More recently, the Labskin model was treated with proinflammatory cytokine interleukin-22 to induce a psoriatic phenotype. The penetration of a psoriatic therapy drug acetretin, into both psoriatic and normal skin equivalents, was observed after 24 and 48 h using MALDI-MSI [32]. MALDI-MSI has several advantages for the analysis of 3D cell cultures and engineered tissue. Many different mediums have been analyzed using this method and there is an abundance of information available on the species observed for each disease type. However, MALDI-MSI does require a matrix application step and does not provide spatial resolution in the very low micrometer range. In general, 3D cell culture models are not large in size. For instance, certain types of single clone tumor spheroids in the current literature are <300 μm in diameter. This produces no difficulty to commonly used techniques such as fluorescent microscopy; however in order to observe the small spheroid microenvironment, higher spatial resolution is required for MSI.

MALDI-MSI is only one of many mass spectrometric imaging methods, all of which come with specific advantages. Laser Ablation Inductively Coupled Plasma Mass Spectrometry (LA-ICP-MS), for example, has also been used in its imaging modality in order to examine the cellular uptake of a second-generation photosensitizer into a tumor spheroid, with and without the assistance of a nanoparticle delivery system. A more homogenous distribution of nanoparticle delivered photosensitizer was determined using imaging, compared to freely dissolved drug [33]. In order to achieve high sensitivity with this system, the drug was tagged with palladium. Having to tag the analyte of interest is a significant disadvantage as, since a target has to be known, it means that the technique is not suitable for de novo discoveries. However, as shown in a more recent study using LA-ICP-MSI, tagging is not always required as long as the analyte of interest has a suitable metal component. Niehoff and colleagues developed a methodology for the imaging of platinum group containing drugs, such as cisplatin, in tumor spheroids, as well as the previously imaged palladium tagged photosensitizer. Quantitation of all the imaged drugs was also achieved by using different matrix-matched standards [34]. In both cases of using LA-ICP-MSI, high spatial resolution was used (<10 μm), which is an advantage of the technique, as well as the need for very little sample preparation. Nevertheless, this method can only be used for de novo discovery when looking at changes in metal groups.

MSI of ex vivo cultured alternative samples is straightforward to develop, as the sample composition is identical to a human tissue biopsy. The differing factor derives from the culture and treatment of the sample outside of the donor. For example, the use of ex vivo human skin for the study of disease and treatment penetration has become more prevalent, possibly due to an increase in sample availability. MALDI-MS, in combination with Ion Mobility Spectroscopy (IMS), has been used to image the endogenous distribution of proteins within human skin. Alternative models, like the in vivo counterparts they represent, are complex, information-rich samples. In this study, the use of Ion Mobility Spectroscopy for additional separation of species of interest aided in the identification of several region-specific peptides [35]. Secondary Ion Mass Spectrometry (SIMS) was also employed for the imaging of ex vivo skin. The aim of this study was to determine the fatty acid enhancing effect on drug penetration into human skin, yet using MSI it was discovered that all fatty acids used in the experiment penetrated the skin; however, only oleic acid demonstrated significant enhanced penetration of the drug [36]. When looking for a high spatial resolution method to observe the alternate model microenvironment in detail, SIMS imaging must be considered. The static SIMS instrumentation is capable of imaging a wide range of samples or a dynamic SIMS instrument could be chosen which possesses the ability to depth profile in order to create three-dimensional images, with some loss of range. SIMS is an imaging method currently capable of single-cell imaging, as shown by Passareli and colleagues who observed the cellular uptake of a pharmaceutical compound in a single macrophage cell [37]. This would provide extremely detailed information of the composition and interaction of cells within a model.

All the methods mentioned so far require the sample to be under vacuum conditions yet there are ambient methods, such as DESI, also potentially capable of alternative model imaging. DESI requires very little sample preparation compared to other methods and can analyze many different sample types, though it is presently used mainly for small molecule and lipid analysis.

Several parameters will need to be considered when analyzing alternative models, as they are both similar to tissue and 2D cell culture models. These parameters will depend on the type of model used, its size, composition, and sample preparation required. For instance, the majority of 3D cell culture models are relatively small, which leads to many of these requiring an embedding step. This additional step also introduces more time to the sample preparation process, which should be considered particularly in the analysis of small molecules. The size of the sample will affect the spatial resolution and ionization method to be used. As there are many alternative models of disease available, sometimes the most appropriate model can be chosen to suit the method plan. For example, the MALDI-MSI study of tumor spheroids has mainly focused on multi-cellular spheroid agglomerates as models, produced using the liquid overlay method, which has been shown to be a less favorable choice for several studies of the tumor microenvironment [38]. Nevertheless, its common use with MALDI-MSI is due to the ease and speed of production of these spheroids, compared to other methods involving growth inside scaffolds, as well as the benefit of the large spheroid size that the method produces (~1 mm). Time required generating the engineered model and ease of production, as well as possibility for integration into high-throughput protocols, are all parameters that require consideration.

However, there are many different alternative models available for the study of disease, many of which are suitable for MSI. As interest in the investigation of these models grows, many other specialized method workflows will emerge.

Published data strongly suggests that MSI will develop into a valuable tool for biomedical research and wider clinical applications. The technique will complement pre-existing biomedical techniques such as immunohistochemistry in both research and in-vitro diagnostics. The use of alternative disease models for MSI within the research environment will produce workflows capable of higher reproducibility, lower cost, higher throughput, and improved ethical methodology. The use of these models in combination with useful methodology will contribute toward the replacement and reduction of the use of animals in research and industry.

Further development of MSI is required to improve suitability to the alternative models available, so that the combined approach can successfully provide the information required in disease characterization and prevention. MSI has been shown capable of providing spatiomolecular information in tumor spheroids, living skin equivalents, and ex vivo human tissues. Due to a considerable interest and scientific effort there are many more designed alternative disease models available which would benefit from the information MSI could provide.

References

1. Cole LM et al (2011) Investigation of protein induction in tumour vascular targeted strategies by MALDI MSI. Methods 54(4):442–453
2. Reis-Filho JS, Lakhani SR (2008) Breast cancer special types: why bother? J Pathol 216(4):394–398
3. Vargo-Gogola T, Rosen JM (2007) Modelling breast cancer: one size does not fit all. Nat Rev Cancer 7(9):659–672
4. Guenther S et al (2015) Spatially resolved metabolic phenotyping of breast cancer by desorption electrospray ionization mass spectrometry. Cancer Res 75(9):1828–1837
5. Mao X et al (2016) Application of imaging mass spectrometry for the molecular diagnosis of human breast tumors. Sci Rep 6:21043
6. Attia AS et al (2012) Monitoring the inflammatory response to infection through the integration of MALDI IMS and MRI. Cell Host Microbe 11(6):664–673
7. Steu S et al (2008) A procedure for tissue freezing and processing applicable to both intra-operative frozen section diagnosis and tissue banking in surgical pathology. Virchows Arch 452(3):305–312
8. Seeley EH, Caprioli RM (2011) MALDI imaging mass spectrometry of human tissue: method challenges and clinical perspectives. Trends Biotechnol 29(3):136–143
9. Fox CH et al (1985) Formaldehyde fixation. J Histochem Cytochem 33(8):845–853
10. Westermark P, Nilsson GT (1984) Demonstration of amyloid protein AA in old museum specimens. Arch Pathol Lab Med 108(3):217–219
11. Battifora H (1986) The multitumor (sausage) tissue block: novel method for immunohistochemical antibody testing. Lab Invest 55(2): 244–248
12. Battifora H, Mehta P (1990) The checkerboard tissue block. An improved multitissue control block. Lab Invest 63(5):722–724
13. Groseclose MR et al (2008) High-throughput proteomic analysis of formalin-fixed paraffin-embedded tissue microarrays using MALDI imaging mass spectrometry. Proteomics 8(18): 3715–3724
14. Djidja M-C et al (2010) Novel molecular tumour classification using MALDI-mass spectrometry imaging of tissue micro-array. Anal Bioanal Chem 397(2):587–601

15. Buck A et al (2015) High-resolution MALDI-FT-ICR MS imaging for the analysis of metabolites from formalin-fixed, paraffin-embedded clinical tissue samples. J Pathol 237(1):123–132
16. Li M, Fan P, Wang Y (2015) Lipidomics in health and diseases - beyond the analysis of lipids. J Glycomics Lipidomics 05(01):126
17. Renshaw S (2007) Immunochemical staining techniques. In: Renshaw S (ed) Immunohistochemistry: methods express. Scion Publishing Ltd, Bloxham, pp 45–96
18. Greek R, Menache A (2013) Systematic reviews of animal models: methodology versus epistemology. Int J Med Sci 10(3):206–221
19. Stranger BE, Stahl EA, Raj T (2011) Progress and promise of genome-wide association studies for human complex trait genetics. Genetics 187(2):367–383
20. Shanks N, Greek R, Greek J (2009) Are animal models predictive for humans? Philos Ethics Humanit Med 4(1):2
21. Mead B et al (2016) Mesenchymal stromal cell-mediated neuroprotection and functional preservation of retinal ganglion cells in a rodent model of glaucoma. Cytotherapy 18(4): 487–496
22. Tamaki C et al (2013) Potentials and limitations of nonclinical safety assessment for predicting clinical adverse drug reactions: correlation analysis of 142 approved drugs in Japan. J Toxicol Sci 38(4):581–598
23. van Meer PJ et al (2012) The ability of animal studies to detect serious post marketing adverse events is limited. Regul Toxicol Pharmacol 64(3):345–349
24. Alnasif N (2014) Penetration of normal, damaged and diseased skin--an in vitro study on dendritic core-multishell nanotransporters. J Control Release 185:45–50
25. Choi SH et al (2014) A three-dimensional human neural cell culture model of Alzheimer's disease. Nature 515(7526):274–278
26. Rodenhizer D et al (2015) A three-dimensional engineered tumour for spatial snapshot analysis of cell metabolism and phenotype in hypoxic gradients. Nat Mater 15:227–234
27. Kappler B et al (2016) The cytoprotective capacity of processed human cardiac extracellular matrix. Materials in. Medicine 27(120): 1–13
28. Li H, Hummon AB (2011) Imaging mass spectrometry of three-dimensional cell culture systems. Anal Chem 83(22):8794–8801
29. LaBonia GJ et al (2016) Drug penetration and metabolism in 3D cell cultures treated in a 3D printed fluidic device: assessment of irinotecan via MALDI imaging mass spectrometry. Proteomics 16(11–12):1814–1821
30. European Commission (2009) Regulation (EC) No 1223/2009 of the European Parliament and of the Council of 30 November 2009 on cosmetic products
31. Mitchell CA et al (2015) Lipid changes within the epidermis of living skin equivalents observed across a time-course by MALDI-MS imaging and profiling. Lipids Health Dis 14(1):84
32. Harvey A et al (2016) MALDI-MSI for the analysis of a 3D tissue-engineered psoriatic skin model. Proteomics 16(11–12):1718–1725
33. Niehoff A-C et al (2014) A palladium label to monitor nanoparticle-assisted drug delivery of a photosensitizer into tumor spheroids by elemental bioimaging. Metallomics 6(1):77–81
34. Niehoff A-C et al (2016) Quantitative bioimaging of platinum group elements in tumor spheroids. Anal Chim Acta 938:106–113
35. Hart PJ et al (2013) Matrix assisted laser desorption ionisation ion mobility separation mass spectrometry imaging of ex-vivo human skin. Int J Ion Mobil Spectrom 16(2):71–83
36. Kezutyte T et al (2013) Studying the penetration of fatty acids into human skin by ex vivo TOF-SIMS imaging. Biointerphases 8(1):3
37. Passarelli MK et al (2015) Single-cell analysis: visualizing pharmaceutical and metabolite uptake in cells with label-free 3D mass spectrometry imaging. Anal Chem 87(13): 6696–6702
38. Fennema E et al (2013) Spheroid culture as a tool for creating 3D complex tissues. Trends Biotechnol 31(2):108–115

INDEX

Laura M. Cole (ed.), *Imaging Mass Spectrometry: Methods and Protocols*, Methods in Molecular Biology, vol. 1618, DOI 10.1007/978-1-4939-7051-3, © Springer Science+Business Media LLC 2017

P

Q

R

S

T

W

Printed by Printforce, the Netherlands